Development of Geopolymer from Pond Ash-Thermal Power Plant Waste

Scrivener Publishing
100 Cummings Center, Suite 541J
Beverly, MA 01915-6106

Publishers at Scrivener
Martin Scrivener (martin@scrivenerpublishing.com)
Phillip Carmical (pcarmical@scrivenerpublishing.com)

Development of Geopolymer from Pond Ash-Thermal Power Plant Waste

Novel Constructional Materials for Civil Engineers

Edited by

Muktikanta Panigrahi

Ratan Indu Ganguly

and

Radha Raman Dash

Scrivener Publishing

WILEY

This edition first published 2023 by John Wiley & Sons, Inc., 111 River Street, Hoboken, NJ 07030, USA and Scrivener Publishing LLC, 100 Cummings Center, Suite 541J, Beverly, MA 01915, USA

For more information about Scrivener publications please visit www.scrivenerpublishing.com.

Wiley Global Headquarters
111 River Street, Hoboken, NJ 07030, USA

For details of our global editorial offices, customer services, and more information about Wiley products visit us at www.wiley.com.

Library of Congress Cataloging-in-Publication Data

ISBN 978-1-394-16652-7

Cover image: Pixabay.Com
Cover design by Russell Richardson

Set in size of 11pt and Minion Pro by Manila Typesetting Company, Makati, Philippines

Printed in the USA

10 9 8 7 6 5 4 3 2 1

This book is dedicated to Professor Basudam Adhikari, the former head of the Materials Science Centre of the Indian Institute of Technology (IIT), Kharagpur, West Bengal, India. On many occasions he extended his support and encouraged our work on new materials. Lab support was also provided to the authors, especially for characterization of materials.

Contents

Preface xi

1 Historical Development of Construction Materials – From Stone Age to Modern Age 1
Ashis Kumar Samal, Muktikanta Panigrahi, Ratan Indu Ganguly and Radha Raman Dash
1.1 Introduction 1
1.2 Chronological Development of Construction 2
1.2.1 Neolithic Age 2
1.2.2 Copper Age and Bronze Age 3
1.2.3 Iron Age and Steel Age 3
1.2.4 Ancient Mesopotamia 4
1.2.5 Ancient Egypt 4
1.2.6 Ancient Greece and Rome 5
1.2.7 Ancient China 8
1.2.8 The Middle Ages 9
1.2.9 The Renaissance 11
1.2.10 The Seventeenth Century 15
1.2.11 The Eighteenth Century 15
1.2.12 The Nineteenth Century 16
1.2.13 The Twentieth Century 17
1.3 Different Types of Ash Used in Construction 18
1.3.1 Wood Ash 19
1.3.2 Rice Husk Ash 19
1.3.3 Cigar Ash 19
1.3.4 Volcanic Ash 19
1.3.5 Quarry Dust 20
1.3.6 Coconut Shell Ash 21
1.3.7 Coal Ash and Fly Ash 21
1.3.8 Fly Ash Generation 24

1.3.9 Nature and Composition of Thermal Power Plant Ashes 24
1.3.10 Pond Ash 29
1.3.11 Various Uses of Pulverized Fuel Ash 31
1.3.12 Importance of Pond Ash Management 32
1.4 Physical Characteristics of Coal Ashes 33
1.5 Coal Ash Utilization 38
1.6 Slag 39
1.6.1 Generation of Slag 40
1.6.2 Slag Properties and Utilization 44
1.7 Geopolymers 45
1.7.1 Constituents of Geopolymers 46
1.7.2 Geopolymer Properties 52
1.8 Durability of Concrete 53
1.9 Accelerated Durability Testing 55
1.10 Conclusion(S) 56
Acknowledgments 56
References 56

2 Fundamentals of Geopolymer Cementitious Materials 71
Muktikanta Panigrahi, Ratan Indu Ganguly and Radha Raman Dash
2.1 Introduction 72
2.2 Parameters of Geopolymer Concrete 78
2.3 Geopolymer Formation Mechanism 78
2.4 Conclusions 81
Acknowledgments 81
References 82

3 Pond Ash (PA)-Based Geopolymer Cementitious Materials 91
Muktikanta Panigrahi, Ratan Indu Ganguly and Radha Raman Dash
3.1 Introduction 92
3.2 Experimental Details 94
3.2.1 Materials 94
3.2.1.1 Pond Ash 94
3.2.1.2 Physical Properties of Pond Ash 96
3.2.1.3 Chemicals 96
3.2.2 Preparation of Geopolymer from Pond Ash 98
3.2.3 Test Methods 100
3.2.4 Results and Discussion 104

3.3 Conclusions 114
Acknowledgments 115
References 116

4 Quantification of Variables on Strength Property of Pond Ash (PA)-Based Geopolymer 123
Muktikanta Panigrahi, Subhasmita Prusty, Ratan Indu Ganguly and Radha Raman Dash
4.1 Introduction 124
4.2 Experimental Details 126
4.2.1 Materials and Method 126
4.2.2 Preparation of Geopolymer from Raw Materials 126
4.2.3 Characterization of Prepared Samples 127
4.3 Results and Discussion 127
4.3.1 Testing of Significance Coefficients 133
4.4 Conclusions 148
Acknowledgments 148
References 149

5 Development of Pond Ash (PA)–High Carbon Ferrochrome (HCFC) Slag-Based Geopolymer Cementitious Materials 151
Muktikanta Panigrahi, Ratan Indu Ganguly and Radha Raman Dash
5.1 Introduction 152
5.2 Experimental Details 156
5.2.1 Source of Materials 156
5.2.2 PA/HCFC Slag-Based Geopolymer (GP) Preparation 157
5.2.3 PA/HCFC-Based Geopolymeric Mortar and Concrete 158
5.2.4 Characterizations of PA/HCFC-Based Geopolymeric Material 158
5.2.5 Results and Discussion 159
5.3 Conclusions 163
Acknowledgments 164
References 164

6 Pond Ash (PA)–Jute Fiber-Based Geopolymer Cementitious Materials 169
Muktikanta Panigrahi, Paresh Biswal, Niharika Patel, Ratan Indu Ganguly and Radha Raman Dash
6.1 Introduction 170
6.2 Experimental Details 175

6.2.1 Chemicals and Materials 175
6.2.1.1 Physical Properties of Jute Fiber 176
6.2.2 PA/Jute Fiber-Based Geopolymer, Mortar and Concrete 178
6.2.3 Results and Discussion 186
6.3 Conclusions 189
Acknowledgments 189
References 189

7 Corrosion of Pond Ash (PA)-Based Geopolymer Products 195
Slipika Panda, Muktikanta Panigrahi, Ratan Indu Ganguly and Radha Raman Dash
7.1 Introduction 196
7.2 Experimental Details 203
7.2.1 Chemicals and Materials 203
7.2.2 Preparation of Pond Ash-Based Geopolymer Products 203
7.2.2.1 Pond Ash-Based Geopolymer Mortar Preparation 203
7.2.2.2 Pond Ash-Based Geopolymer Concrete Preparation 204
7.2.3 Characterizations of Pa-Based Geopolymer GP Mortar/Concrete (Before and After) Corrosion 206
7.2.4 Results and Discussion 207
7.3 Conclusions 220
Acknowledgments 220
References 221

8 Applications, Challenges and Opportunities of Geopolymer Materials 227
Ashis Kumar Samal, Muktikanta Panigrahi, Ratan Indu Ganguly and Radha Raman Dash
8.1 Introduction 228
8.2 Challenges 234
8.3 Opportunity 234
8.4 Conclusions 235
Acknowledgments 235
References 235

Index 241

Preface

The idea of this book originated with a book chapter published in 2014* that extensively reviewed the development of geopolymer from fly ash (FA). Simultaneously, in a program sponsored and financed by M/S, NALCO Company, work was going on in the production of ceramic wall tiles from fly ash obtained from thermal power stations. While working on the project, industry management discussed their bottom ash disposal problem, which they used to dump in adjoining ponds. Therefore, the authors examined the composition of bottom ash vis-a-vis pond ash. Since the main ingredients are silica and alumina, it was thought to use this material for making geopolymer.

Some experiments were performed to determine the feasibility of developing geopolymer from pond ash (PA). With encouraging experimental results, the authors submitted a proposal to the Ministry of Mines, Government of India. The project was sanctioned and work started in full swing, with reports periodically submitted to the Ministry of Mines, Government of India.

The authors observed that utilization of waste materials, such as pond ash, is a global challenge since it endangers the environment. Presently, R&D is being carried out to utilize these materials for producing value-added products. In the present investigation, an effort has been made to utilize fly ash/pond ash (waste materials from thermal power plants) for producing a novel material called geopolymer. Red mud, slags, etc., were mixed with fly ash to produce geopolymer with enhanced strength.

Geopolymer (GP) can replace cement, as shown by a few European countries, and some permanent structures constructed with GP are now appearing in a few advanced countries. Geopolymer and geopolymer concrete is thought to be suitable for construction of roads, buildings, etc., and will eventually fully or partially replace cement. Thus, it is no wonder why some scientists are trying to develop GP using waste materials.

This book highlights the formation mechanism of GP from pond ash. This will perhaps be the first attempt to use pond ash for developing GP.

The properties of structures made with GP concrete are found to be comparable with those made with cement concrete. Systematic investigations have been carried out in order to understand the chemistry or GP formation with pond ash materials. Performances of these materials above ambient temperature, as well as under different environments, are evaluated. Results indicate the possible replacement of cement with newly invented GP prepared from pond ash.

Muktikanta Panigrahi
Ratan Indu Ganguly
Radha Raman Dash
June 2022

[*] Pradeep Kumar Rana, Radha Raman Dash and Ratan Indu Ganguly, "Geopolymer from Industrial Wastes" in *Advanced Composites for Aerospace, Marine, and Land Applications*, Tomoko Sano, T.S. Srivatsan, and Michael W Peretti (eds.), TMS (The Minerals, Metals & Materials Society), 2014.

1

Historical Development of Construction Materials – From Stone Age to Modern Age

Ashis Kumar Samal[1], Muktikanta Panigrahi[2]*, Ratan Indu Ganguly[3] and Radha Raman Dash[4]

[1]*Department of Civil Engineering, Gandhi Institute of Engineering and Technology University, Gunupur, India*
[2]*PG Department Materials Science, Maharaja Sriram Chandra Bhanja Deo University, Keonjhar Campus, Odisha, India*
[3]*Department of Metallurgical Engineering, National Institute of Technology, Raurkela, Odisha, India*
[4]*CSIR-National Metallurgical Laboratory, Jamshedpur, Jharkhand, India*

Abstract
In this chapter, the history of construction and the synthesis, spectroscopic characteristics, thermal and mechanical behavior of construction materials, such as their compressive strength, and the durability of industrial waste are presented. Because of their potential application in construction sectors, the mechanical and corrosion properties of the materials are studied in detail. In addition to ancient construction materials, different industrial waste, such as fly ash, pond ash, bottom ash, various slags, and geopolymers are also discussed.

Keywords: Ancient construction materials, thermal power plant wastes, non-ferrous and ferrous industry waste geopolymer, physico-mechanical properties, geopolymer synthesis, geopolymer properties-mechanical and corrosion

1.1 Introduction

The history of construction is connected with many other fields of structural engineering. An investigation into how builders lived and their

**Corresponding author*: muktikanta2@gmail.com

Muktikanta Panigrahi, Ratan Indu Ganguly and Radha Raman Dash (eds.) Development of Geopolymer from Pond Ash-Thermal Power Plant Waste: Novel Constructional Materials for Civil Engineers, (1–70) © 2023 Scrivener Publishing LLC

activities has been recorded, which allows us to analyze constructed buildings and other structures, the tools used to construct them, and different uses of building materials. The history of building has evolved over time, along with the key principles of durability of the materials used, increasing the height and span of structures, degree of control exercised over the interior environment, and finally the energy available for the construction process [1, 2].

1.2 Chronological Development of Construction

1.2.1 Neolithic Age

The Neolithic Age is also known as the New Stone Age, the time period from roughly 9000 BC to 5000 BC. The tools used at that time were made from natural materials such as bone, hides, stone, wood, grasses, and animal fibers. People used tools, such as axes, choppers, adzes, and celts, to cut by hand. Also, to scrape, chop such as with a flake tool, pound, pierce, roll, pull and leaver. Building materials included bones such as mammoth ribs, hides, stone, metal, bark, bamboo, clay, lime plaster, and more. Neolithic tools used in the reconstruction of a pithouse-type dwelling made with mammoth bones are shown in Figure 1.1. Dwellings at the time, such as caves and rock shelters, were very simple, and resembled tents like the Inuit's tupiq and huts sometimes built as pithouses. These constructed structures are found in the stone-built Neolithic settlement of Skara Brae in Scotland and in Europe's Neolithic village. In the Neolithic period, mold-made mud bricks were first used for the structures found in Jericho [3–5].

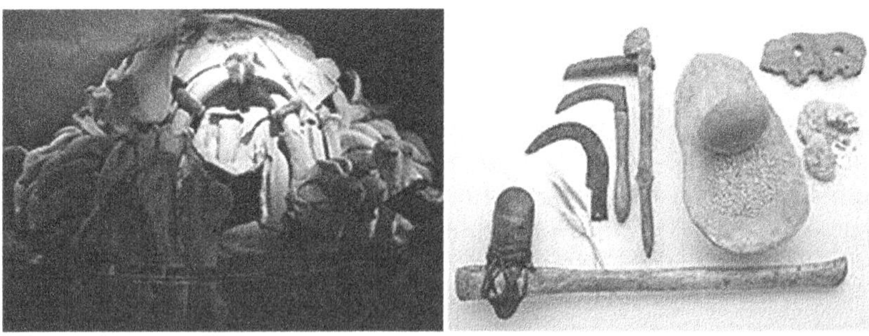

(a) (b)

Figure 1.1 Reconstruction of a pit-house type of dwelling made with mammoth bones (a) and Neolithic tools (b) [1].

Neolithic architecture ranges from the tent to the megalith, i.e., arrangement of large stones. Temples, tombs, and dwellings were rock-cut architecture. The most remarkable Neolithic structure, i.e., iconic megalith, in Western Europe is Stonehenge.

1.2.2 Copper Age and Bronze Age

Copper started being used prior to 5,000 BC, and bronze around 3,100 BC. Copper and bronze were used to make tools such as axes and chisels, and were also used in the cutting edge tools of the time [1]. Figure 1.2 shows a bronze saw developed during this period that was found at the archaeological site of Akrotiri that can be seen in the Museum of Prehistoric Thera in Santorini, Greece.

In the Bronze Age, the corbelled arch came into use (beehive tombs). Prior to the wheel, heavy loads were moved on boats, sledges (a primitive sled) or on rollers.

The Egyptians began building stone temples using the post and lintel construction method. The Greeks and Romans used similar techniques [1].

1.2.3 Iron Age and Steel Age

The Iron Age is a cultural period from about 1200 BC to 50 BC. During this period, iron was widely used to make tools and weapons. After 300 BC, steel was produced by mixing carbon and iron, ushering in the Steel Age.

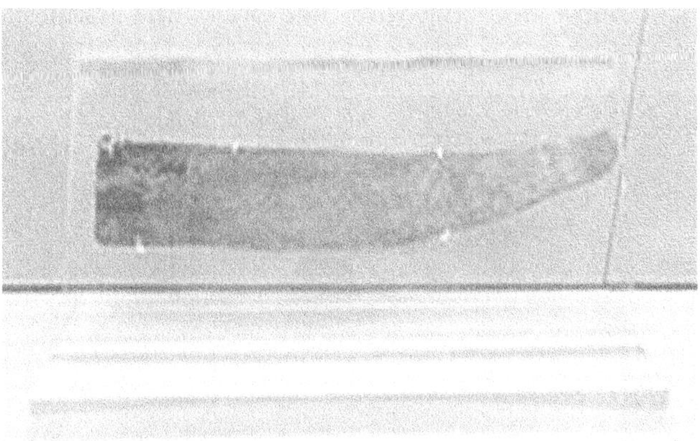

Figure 1.2 A bronze saw from the archaeological site of Akrotiri on display at the Museum of Prehistoric Thera in Santorini, Greece.

Steel can be hardened and tempered producing a sharp, durable cutting edge. A new woodworking tool allowed by the use of steel is the hand-plane [1].

1.2.4 Ancient Mesopotamia

Evidence survives of the large-scale buildings built by the ancient Mesopotamians. The smaller dwellings only survive in traces of foundations. The major technical achievements of Mesopotamian construction are great cities such as Uruk and Ur. Out of these, the Ziggurat of Ur is an outstanding building of the period. Another fine construction is the ziggurat at Chogha Zanbil in modern Iran. The chief building material was the mud-brick prepared by wooden molds with varied size, including rectangular and square-sized bricks. By 3500 BC, fired bricks and stone were used for preparing pavement. The later Mesopotamians, particularly in Babylon and thence Susa, developed glazed brickwork of a high quality for decorating the interiors and exteriors of their buildings [6].

1.2.5 Ancient Egypt

The Egyptian pharaohs built huge structures in stone. The arid climate has preserved much of their ancient buildings. In ancient Egypt, adobe (i.e., sunbaked mud brick) was used for constructing ancillary buildings and normal houses. The characteristics of mud-brick, which allows rain to drain quickly away, are ideal for the hot, dry climate of Egypt. The Ramesseum in Thebes, Egypt (Luxor), is constructed by mud brick. Also, extensive storehouses are built with mud-brick. The grandest buildings,

(a) (b)

Figure 1.3 (a) Great Pyramid of Giza and (b) the Menkaure Pyramid [1].

such as the pyramids (Figure 1.3) and temples are constructed in stone, which are massive masonry blocks.

The Egyptians did their work using relatively primitive techniques. They transported massive stones for long distances by rollers, ropes and sledges hauled by large numbers of workers. They invented the ramp, lever, lathe, oven, ship, paper, irrigation system, window, awning, door, glass, a form of plaster of Paris, the bath, lock, shadoof, weaving, a standardized measurement system, geometry, silo, a method of drilling stone, saw, steam power, proportional scale drawings, enameling, veneer, plywood, rope truss, and more. They also knew how to lift stones to great heights and erect obelisks. Most theories center on the use of ramps. Imhotep, who lived circa 2650–2600 BC, is acknowledged as the first recorded architect and engineer.

1.2.6 Ancient Greece and Rome

The ancient Greeks tended to build most of their common buildings with mud bricks. Some of their most dramatic structures are the Greek

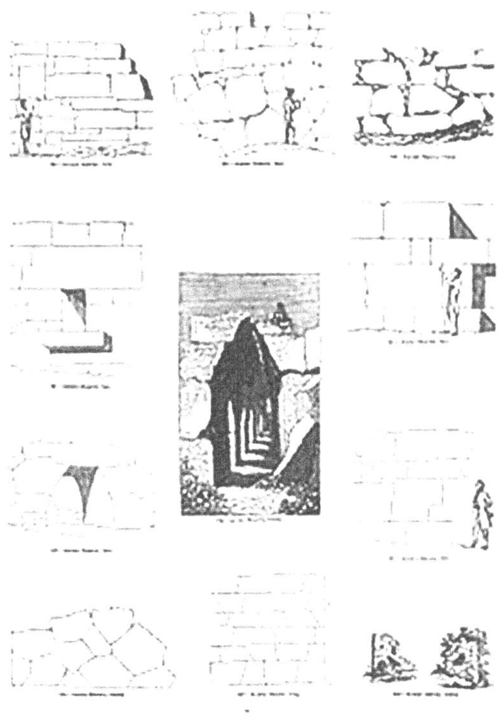

Figure 1.4 Various masonry techniques of ancient Greece and Rome [1].

temples [7]. Figure 1.4 shows various masonry techniques used in ancient Greece and Rome.

They used many advances in technology, i.e., plumbing, spiral staircase, central heating, urban planning, the water wheel, crane, etc. The oldest construction drawing is in the Temple of Apollo at Didyma. The spans are very simple beam and post structures spanning stone walls. For the longer spans, it is not known whether the Greeks or Romans invented the truss. Prior to 650 B.C.E., the famous ancient Greek temples were built of wood, and after this date, stone was used to build temples [8]. Fired clays were used for decorations. Next, fired bricks with lime mortar started being employed. Very prominent buildings were roofed in stone tiles that mimicked the form of their terracotta counterparts.

Because the process used by later cultures of constructing a limited number of buildings using thin skins of finished stones over rubble cores was a slow, expensive and laborious process, the Greeks used large cut blocks joined with metal cramps. Building structures were mostly a simple beam and column system without vaults or arches. Also, they constructed some groin vaults, arch bridges, and made the Lighthouse of Alexandria with the Egyptians (one of the Seven Wonders of the Ancient World), which was the first high rise. Their surveying skills enabled them to set out the incredibly exact optical corrections of buildings like the Parthenon [9].

The structure of buildings made by the Romans, such as the Pantheon in Rome, are very well preserved. The great Roman development in building materials is Roman cement, which is a hydraulic lime mortar which would harden under water. It is a strong material used for bulk walling. It was also the Romans who developed the treadwheel crane shown in Figure 1.5. They are used brick or stone to build the outer skins of the wall and also filled the cavity with massive amounts of concrete. Later, they used wooden shuttering, which was removed for the concrete to cure. A temple is made of Roman concrete in the 1st century BC is the Temple of Vesta in Tivoli, Italy. The concrete was made from rubble and mortar. It was cheap because of the low production cost. They used it to make arches, barrel vaults and domes, spans. They also developed systems of hollow pots to make domes and sophisticated heating and ventilation systems for their thermal baths. The Romans also made bronze roof tiles [10] and during the Iron Age in Germany (1st to 3rd century AD) they developed the hand plane shown in Figure 1.6.

The Romans used lead as a roof covering material. They also used glass in construction; colored glass for mosaics and clear glass for windows [9]. The Romans also invented central heating with a hypocaust system that circulated the exhaust of a wood or coal fire.

Figure 1.5 Roman treadwheel crane at Bonn, Germany [1].

Figure 1.6 Roman hand plane developed during the Iron Age in Germany (1st to 3rd century AD) [1].

Most construction was done by slaves or free men. The use of slave labor definitely cut costs, which was one of the reasons for the scale of some of the structures. The construction of very large structures could only be undertaken with vast numbers of workers.

The Romans invented the waterwheel, sawmill, and arch. After 100 CE, they also used glass for architectural purposes. Also, they used double glazing, which acted as insulated glazing. Moreover, they made roads such as corduroy roads and paved roads, sometimes supported on raft or pile foundations and bridges. The Romans also developed sophisticated timber cranes allowing them to lift considerable weights (up to 100 tonnes) to great heights. Trajan's column in Rome contains some of the largest stones. A list of the longest, highest and deepest Roman structures can be found in ancient architectural records. Roman building inventiveness extended over bridges, aqueducts, and covered amphitheaters. Their sewerage and water-supply works were remarkable. However, very little evidence has survived concerning the form of their timber roof structures. Possibly, the triangulated roof trusses built by Romans did not exceed a span of 30 meters.

1.2.7 Ancient China

China is a cultural hearth area of eastern Asia, and many building methods evolved from China. The famous Great Wall of China provides an example of their construction methods. Constructed between the 7th and 2nd centuries BC, it was built with rammed earth, stones, and wood; and later bricks and tiles with lime mortar. They used wooden gates for blocking passageways. The oldest archaeological examples of mortise and woodworking joints, dating back to about 5000 BC, are found in China. The Yingzao Fashi is the oldest technical manual on Chinese architecture. Since the Chinese followed the state rules for thousands of years, many of the ancient buildings built using these methods and materials were still used in the 11th century. Chinese temples were usually made with wooden timber frames on an earth and stone base. The oldest wooden building is the Nanchan Temple (Wutai) dating back to 782 AD. Since the temple builders regularly reconstructed the wooden temples, some parts are of different ages. Generally, traditional Chinese timber frames do not use trusses. The Songyue Pagoda is the oldest brick pagoda dating to 523 AD. It was built with yellow fired bricks and clay mortar, and has twelve sides and fifteen levels of roofs. Built in 595–605 AD, the Anji Bridge is the world's oldest open-spandrel stone segmental arch bridge. It was built with sandstone and joined with dovetail, iron joints. Most of the Great Wall restored sections were built with bricks and cut stone blocks/slabs. If bricks and blocks were not available, the builders used local materials such as tamped earth, uncut stones, wood, and even reeds. Wood was used for forts as an auxiliary material.

In mountain areas, workers excavated stone to build the Great Wall. Using the mountains as footings, the outer layer of the Great Wall was built with stone blocks (and bricks), and filled with uncut stone and anything else available.

On the plains, workers rammed compact layers of local soil (sand, loess, etc.) into the Great Wall. Jiayuguan's Great Wall section in west China was mainly built with dusty loess soil.

Sand was used as a fill material between reed and willow layers. In desert areas, builders made use of reeds and willow. Jade Gate Pass (Yumenguan) Great Wall Fort was constructed with 20-cm layers of sand and reed at an impressive 9 meters high.

Bricks with lime mortar were mostly used to build the Ming Dynasty Great Wall. Workers built brick and cement factories using local materials near the wall.

1.2.8 The Middle Ages

The Middle Ages of Europe is the period that extends from the 5th to 15th centuries AD that falls between the Western Roman Empire and the Renaissance. It is divided into Pre-Romanesque and Romanesque periods.

In the Middle Ages, the Europeans used some Roman techniques to construct the fortifications, castles and cathedrals. These techniques, including iron ring-beams, appear to have been used in the Palatine Chapel at Aachen in 800 AD, where it is believed builders from the Longobard kingdom in northern Italy contributed to the work [10]. The 9th century saw a revival of stone buildings and the Romanesque style of architecture began in the late 11th century. Also of note are the famous stave churches found in Scandinavia [11]. In the 13th century, Villard de Honnecourt recorded details of buildings of the Gothic era. Some of his drawings have survived, including one of the flying buttresses of Rheims Cathedral at Reims, ca. AD 1320–1335, shown in Figure 1.7.

Most buildings in Northern Europe were built of timber until about 1000 AD, whereas adobe remained predominant in Southern Europe. Brick continued to be manufactured in Italy from 600–1000 AD. The skill of brick-making had largely disappeared along with that for burning tiles. Roofs were mainly thatched. Houses were small and gathered around a large communal hall. Monasticism spread more sophisticated building techniques. The Cistercians may have been responsible for introducing brick-making technology to different areas—from the Netherlands, through Denmark and Northern Germany to Poland—leading to Brick Gothic. Brick remained the most popular prestige material in these areas throughout the period. Everywhere else, buildings were typically in timber or stone. Medieval stone walls were built using cut blocks on the outside of

Figure 1.7 Villard de Honnecourt's drawing of the flying buttresses of Reims Cathedral, ca. AD 1320–1335 [1].

the walls and rubble infill with weak lime mortars. Due to the poor hardening behavior of these mortars, the settlement of the rubble filling was cause for concern, and continues to be a major problem. Workers transported large stones on ox-drawn sledges to construct churches, as depicted in a sculpture from the 10th century Korogho church in Georgia shown in Figure 1.8.

There were no standard textbooks on building in the Middle Ages. Master craftsmen transferred their knowledge through apprenticeships and from father to son. Trade secrets were closely guarded, as they were the source of a craftsman's livelihood. Drawings only survive from the later period. Parchment was too expensive to be commonly used and paper did not appear until the end of the period. Models were used for designing structures and could be built to large scales. Details were mostly designed at full size on tracing floors, some of which survive.

Figure 1.8 Workers transport a large stone on an ox-drawn sledge for the construction of a church. A sculpture from the 10th century Korogho church in Georgia [1].

At this time, buildings were built by paid workers. Unskilled work was done by laborers paid by the day and skilled craftsmen served apprenticeships or learned their trade from their parents. It is unclear whether there were women members in the guilds that monopolized a particular trade in a particular area. A built town was very small and was dominated by the homes of a small number of rich nobles or merchants and by cathedrals/churches.

In the period of 600–1100 AD, Romanesque buildings were entirely roofed in timber or had stone barrel vaults covered by timber roofs. The Gothic style of architecture developed in the twelfth century had vaults, flying buttresses and pointed gothic arches achieved in stone. Thin stone vaults and towering buildings were built using the trial-and-error method. The pile driver was invented around 1500.

In the Middle Ages, the scale of fortifications and castle building was impressive. The outstanding buildings of the period, such as Beauvais Cathedral, Chartres Cathedral, King's College Chapel and Notre Dame, Paris, were Gothic cathedrals with thin masonry vaults and walls of glass.

1.2.9 The Renaissance

In Italy, the invention of moveable type and the printing press during the Renaissance changed the character of building. The rediscovery of

Vitruvius had a strong influence. In the Middle Ages, buildings were designed by those who built them. The master mason and master carpenters learned their skills by word of mouth and relied on experience, models and the rule of thumb, which is to determine the sizes of building elements. However, Vitruvius describes the education of the perfect architect in detail. They must be skilled in all the subjects of arts and sciences disciplines. Filippo Brunelleschi was the first Renaissance style architect. He started life as a goldsmith and educated himself in Roman architecture by studying the ruins. He went on to engineer the dome of Santa Maria del Fiore in Florence.

In this period, the major breakthroughs were related to the technology of conversion. In western Europe, water mills were used to saw timber and convert it into planks. Bricks were also used in huge quantities. In Italy, brick makers organized into associations. They used kilns mostly in rural areas because of the risk of fire and the easy availability of firewood materials. Brick makers were paid by the brick, which gave them an incentive to make them too small. As a result, legislation was laid down regulating the minimum sizes. Each town kept measures against which bricks were to be compared. There was an increasing demand for ironwork to be used in roof carpentry for straps and tension members. The iron was fixed using forelock bolts. Screw-threaded bolts and nuts made during this period could, for example, be observed in the clocks of this period. However, since making them was labor-intensive, they could not be used in big structures. Roofing was usually terracotta roof tiles. The Italians followed Roman precedents. In northern Europe, people used plain tiles and stone remained the material of choice for prestige buildings. A church in Kizhi, Russia, listed as a UNESCO World Heritage Site building constructed entirely out of wood using the log building technique is shown in Figure 1.9.

During the Renaissance, the rebirth of classical culture was reflected by architects, drastically transforming building design. Prior to this, architecture was viewed as a technical art that required an artisan. The change in architecture and the architect are key to understanding the changes in the design process. The Renaissance architect was often an artist, i.e., a painter or sculptor, who had to provide detailed drawings to the craftsmen. This process involved he ability to make the drawing and the intellectual capacity to invent the design. The architect was only infrequently involved in particularly difficult technical problems since the technical part of architecture was mainly left up to the craftsmen. This changed how buildings were designed. Whereas the Medieval craftsmen were inclined to approach a problem with a technical solution, the Renaissance architects started with

Figure 1.9 Listed as a UNESCO World Heritage Site, this church in Kizhi, Russia, is constructed entirely out of wood using the log building technique [1].

an idea and then searched around for a way of making it work. This led to extraordinary progress in engineering.

Labor in the Renaissance was the same as in the Middle Ages. Buildings were constructed by paid workers. Unskilled workers were taken on as day laborers and paid accordingly. Apprentices studied under the guidance of a master artist or parents of apprentices signed a contract with the master that set out the terms of the training. Crafts and professions were governed by guilds holding a monopoly on a particular trade in a particular area. Towns were very small by modern standards and were dominated by the homes of a small number of rich nobles or merchants and by cathedrals and churches.

The return to classical architecture created problems for the Renaissance buildings. Since the builders did not use concrete, comparable vaults and domes had to be recreated in brick or stone. The greatest technical feats were undoubtedly in these areas. The first major breakthrough was the dome of Santa Maria del Fiore, in which Brunelleschi managed to invent a way of building a huge dome without formwork by using the weight and placement of bricks to keep them in position and the shape of the dome to

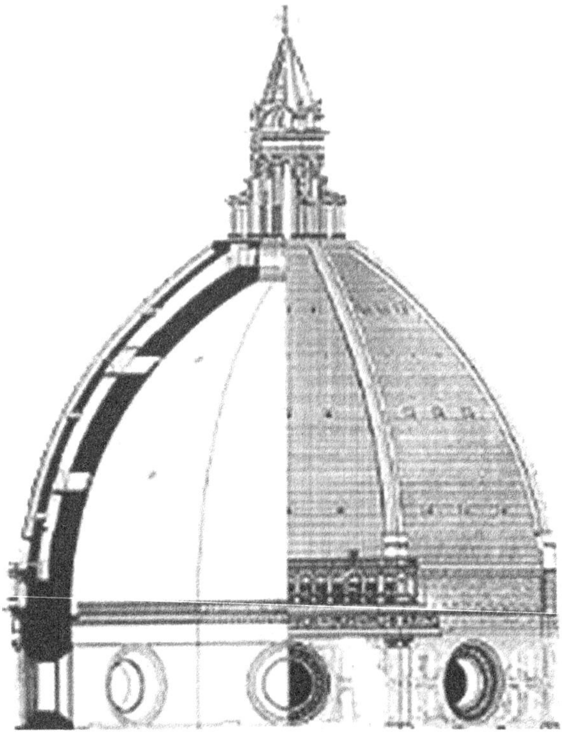

Figure 1.10 The structure of the dome of the Florence cathedral showing the double-skin structure [1].

Figure 1.11 Pieter Bruegel the Elder's Tower of Babel, illustrating construction techniques of the 16th century [1].

keep it standing. The exact way in which the dome was constructed is still a subject of debate today. In an attempt to deal with hoop stresses, the double-skinned dome is linked by ribs and has a series of wooden and stone chains around it at intervals.

Completed in 1446, the size of Brunelleschi's dome was surpassed by that of St Peter's, which was built using flying scaffolding that is supported on the cornices and constructed using two stone shells. The double-skin structure of Brunelleschi's dome of Florence cathedral is shown in Figure 1.10.

Pieter Bruegel the Elder illustrated the construction techniques of the 16th century in the painting shown in Figure 1.11.

1.2.10 The Seventeenth Century

The seventeenth century saw the emergence of modern science. Major breakthroughs in building construction towards the end of the century engendered the use of experimental science by architects and engineers in their buildings. However, in the seventeenth century, architects and engineers still strongly relied on experience, rule of thumb and scale models.

In this period, Iron was progressively used in structures. For example, iron rods were used to repair Salisbury Cathedral and strengthen the dome of St. Paul's Cathedral. Most buildings had stone ashlar surfaces covering rubble cores held together with lime mortar; and experimental use of lime with other materials was used to provide a hydraulic mortar. In England, France and the Dutch Republic, cut and gauged brickwork was used in ornate facades. The triangulated roof truss used by Inigo Jones and Christopher Wren was also introduced to England in this period.

In this period, the construction method remained largely medieval even after the discovery of experimental science. Even though flying scaffolds were used in St. Paul's Cathedral, England, and in the dome of St. Peters, Rome, the same type of timber scaffolding used for centuries was employed. The use of cranes and scaffolding mainly depended on timber. Complex systems of pulleys allowed lifting loads large loads, and long ramps were used to drag loads up to the upper parts of buildings.

1.2.11 The Eighteenth Century

In the eighteenth century, architects and engineers became more and more professionalized. Increasingly sophisticated experimental science and mathematical methods were used in buildings. At the same time, the industrial revolution contributed to the size of cites, and the pace and quantity of construction gradually increased. In this period, the main breakthroughs were the use of cast iron and wrought iron. Iron columns were used in Wren's designs for the House of Commons and several churches in London. These columns also supported the galleries. In the second half of the eighteenth century, the cost of iron production was reduced, allowing its use in the construction of major pieces of iron engineering.

The Iron Bridge at Coalbrookdale is a particularly good example of major iron pieces. Large-scale mill construction needed fire-proof buildings; therefore, cast iron was progressively used for columns and beams. An early example of wrought-iron used in construction is the roof of the Louvre in Paris.

Even though steel was used in the manufacture of different tools in this period, it could not be made in the quantities needed for building; therefore, brick production progressively increased. Although bricks were used to build many buildings in Europe, they were often coated in lime render or patterned to look like stone. Brick production changed little. Bricks were molded by hand and fired in kilns, which was no different from the production method in previous centuries. Terracotta in the form of Coade stone was used as artificial stone in the UK.

1.2.12 The Nineteenth Century

The manifestation of the industrial revolution took place in the nineteenth century in new transportation installations, such as railways, canals and macadam roads, which required a large financial investment. New construction devices, such as steam engines, machine tools, explosives and optical surveying, came into being. The technological advancements of the steam engine combined with the circular saw and machine cut nails led to the use of balloon framing. Due to the progressive use of these new technologies, traditional timber framing was reduced [12]. An example of the construction that took place during this time is the Woolworth Building, which is shown under construction in 1912 in Figure 1.12.

Mass production of steel was feasible in the mid-19th century and was used in I-beams and reinforced concrete. Also, glass panes went into mass production, and went from being a luxury product to being available to everyone. Other improvements of the time included plumbing that provided common access to drinking water and sewage collection; and application of building codes, particularly for fire safety.

Figure 1.12 Woolworth Building under construction in 1912 [1].

1.2.13 The Twentieth Century

In the early 20th century, the elevators and cranes of the second industrial revolution made high rise buildings and skyscrapers possible. Heavy equipment and power tools reduced the need for manpower. Additionally, the new technologies of prefabrication and computer-aided design emerged. Trade unions formed to protect the health and well-being of construction workers by enforcing occupational safety. To this end, the use of protective equipment, such as hard hats and earmuffs, were essential at most construction sites. In the 20th century, governmental construction projects were used as part of macroeconomic stimulation policies. For economy of scale, infrastructure was planned for entire suburbs, towns and cities, and constructed within the same project. At the end of the 20th century, ecology, energy conservation and sustainable development started to become more important construction issues. By the end of the twentieth century, steel and concrete construction were themselves becoming the subject of historical investigation [13–19].

1.3 Different Types of Ash Used in Construction

Ash/ashes is the solid remnants of fires. Specifically, it refers to all non-aqueous, non-gaseous residues. After chemical analysis of the mineral and metal content of chemical samples, ash is the non-gaseous, non-liquid residue after complete combustion. It usually contains an amount of combustible organic or other oxidizable residues. There are various types of ashes in our environment, as shown in Figure 1.13 and described below.

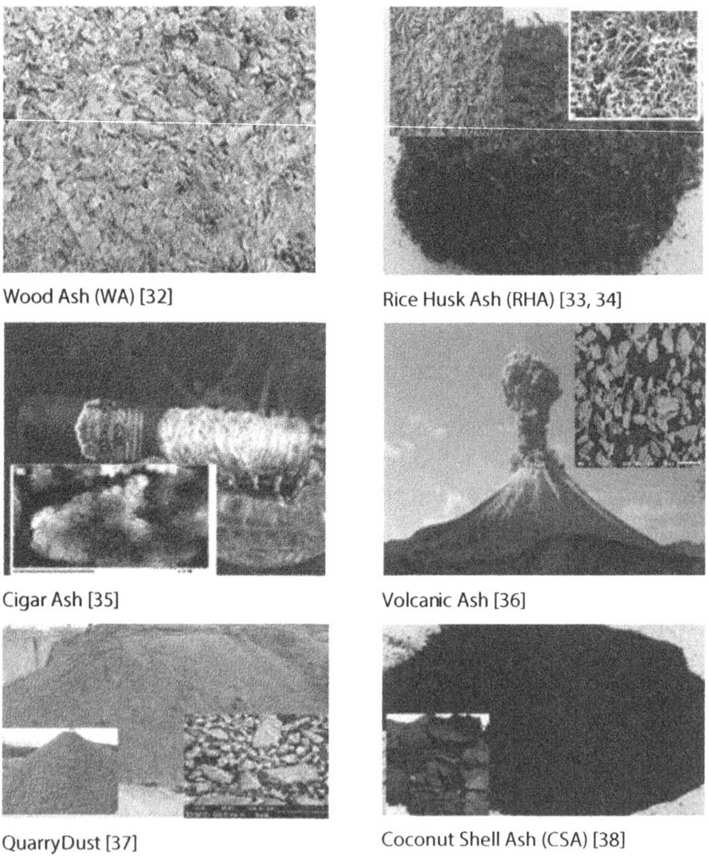

Figure 1.13 Different type of solid ashes.

1.3.1 Wood Ash

Wood ash is one of the ashes in the ash family. It is the powdery residue remaining after the combustion of wood. Gardeners usually use it as a good source of potash to ameliorate soil. There have been many reports on the chemical composition of wood ash, according to which calcium carbonate ($CaCO_3$) is the chief constituent at temperatures below 750 °C [20]. At temperatures above 750 °C, calcium oxide (CaO) is the major constituent [21]. Other constituents in wood ash are Fe, Si, Al, Mn, As, Cd, Pb, Cr, Ni, and V.

1.3.2 Rice Husk Ash

Rice hulls are the hard protective coverings of grains of rice. They are used in various applications such as building material, fertilizer, insulation material, or fuel. Rice hulls are part of the chaff of rice. The hull is hard to eat or swallow. It is mostly indigestible to humans due to its enriched fiber components. The residue after combustion of rice hulls is called rice husk ash (RHA).

The ash is a potential source of reactive amorphous silica. It has a variety of applications in materials science. Mostly, RHA is used in the production of Portland cement [22], which is finer than cement. Silica is the basic component of sand. That is why RHA is used with cement for plastering and concreting. It offers compactness in concrete. Also, the ash is a very good thermal insulation material. Because of its fineness, it is a very good candidate for sealing fine cracks in civil structures. Rice husk ash has long been used in ceramic glazes in rice growing regions of China and Japan [23]. Because of its 95% silica content, it is an easy way of introducing the necessary silica into the glaze.

1.3.3 Cigar Ash

Cigar ash is the ash produced by a cigar as it is smoked. Cigar ash is mixed with chalk to make a dentifrice or tooth powder. Also, it is mixed with poppy seed oil to prepare paint in shades of grey. Generally, during smoking, the ash is an unwanted product which is to be disposed of.

1.3.4 Volcanic Ash

Volcanic ash consists of fragments of rock, minerals, and volcanic glass, created. It is during volcanic eruptions. These fragments are generally very small, measuring less than 2 mm (0.079 inches) in diameter [24]. Volcanic ash is formed during explosive volcanic eruptions when dissolved gases in magma expand and escape violently into the atmosphere. The force of the gasses shatters the magma and propels it into the atmosphere where it solidifies into fragments of volcanic rock and glass. Ash is also produced when magma comes into contact with water, causing the shattering of magma. Ash is transported by wind up to thousands of kilometers away.

Due to its wide dispersal, the ash can have a number of negative impacts on society, viz. animal and human health, disruption to aviation, disruption to critical infrastructure (e.g., electric power supply systems, telecommunications, water and wastewater networks, transportation), primary industries (e.g., agriculture), buildings and structures.

Explosive eruptions arise when magma decompresses. It allows dissolving volatiles and is resolved into gas bubbles [25]. As more bubbles nucleate, foam is produced. When fragmentation occurs, violently expanding bubbles tear the magma apart into fragments, which are ejected into the atmosphere where they solidify into ash particles. Fragmentation is a very efficient process of ash formation and is capable of generating very fine ash even without the addition of water [26].

Volcanic ash is also made through phreatomagmatic eruptions. During these eruptions, fragmentation occurs when magma comes into contact with bodies of water (such as the sea, lakes and marshes) groundwater, snow or ice. As the magma, which is significantly hotter than the boiling point of water, comes into contact with water, an insulating vapor film forms [27]. Ultimately, this vapor film will collapse. This leads to direct coupling of the cold water and hot magma. This helps to increase the heat transfer rate and leads to the rapid expansion of water and fragmentation of the magma into small particles. These are subsequently ejected from the volcanic vent.

1.3.5 Quarry Dust

Quarry dust is a byproduct of the crushing process. It is a concentrated material to use as fine aggregates. In quarrying activities, a rock is crushed into various sizes; during the process the dust generated is called quarry dust and it is formed as waste. Therefore, the dust becomes a useless material. Hence, quarry dust is used for different purposes in the construction industry, i.e., as building materials, road development materials, aggregates, bricks, and tiles [28].

1.3.6 Coconut Shell Ash

Coconut shell, one of the most natural fillers, is manufactured in different countries such as Malaysia, Indonesia, Thailand, and Sri Lanka. Many researchers are devoted to its use as a natural filler for composite preparation. Therefore, coconut shell filler is a potential candidate for the development of new composites. This is due to its high strength and modulus properties. Because of its high lignin content, it makes composites more weather resistant. Hence, it is more suitable for application in construction sectors. Coconut shell flour is also widely used to make furnishing materials, rope, etc. Also, coconut shells absorb less moisture. This is due to its low cellulose content [29].

Many researchers have put effort into preparing carbon black from different agricultural by-products such as coconut shell, apricot stones, sugarcane bagasse, nutshells, forest residues and tobacco stems. Coconut shells have little or no economic value and their disposal may cause environmental problems. Coconut shells can be converted into activated carbon for use as an adsorbent for purifying water or industrial and municipal effluents, which is an added value to these agricultural commodities. It also helps to reduce the cost of waste disposal. Therefore, it affects the cost of existing commercial carbons [30, 31].

1.3.7 Coal Ash and Fly Ash

As a crucial component of growth and development, energy has become the driving force of the modern economy. As can be seen in Figure 1.14, fossil fuels are the main energy source. Oil, one of the energy sources, makes up 43% of the total energy demand (Figure 1.15). Other sources are coal (27%) and natural gas (15%).

Coal is one of the most abundant low-cost fossil fuels found in the Earth's crust. The largest coal deposits are found in China, India and the United States. In the last half century, coal has played a key role in the generation of energy worldwide.

China is the world's largest consumer of coal, and is predicted to use half the global coal by 2035 (Figure 1.16). India shows the largest growth in coal consumption (435 M ton), two-third of which is mostly used in its power sector; whereas the United States has become the world's second largest consumer of coal.

During the last few decades, there was a dramatic increase in coal ash production in the world. This was due to the increased amounts of energy from coal-thermal power plants. Researchers [7–9] have assembled data

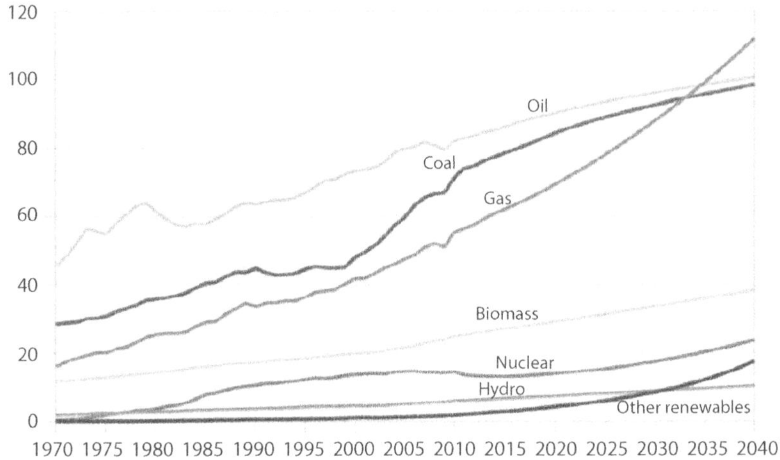

Figure 1.14 Global energy mix by fuel type [39].

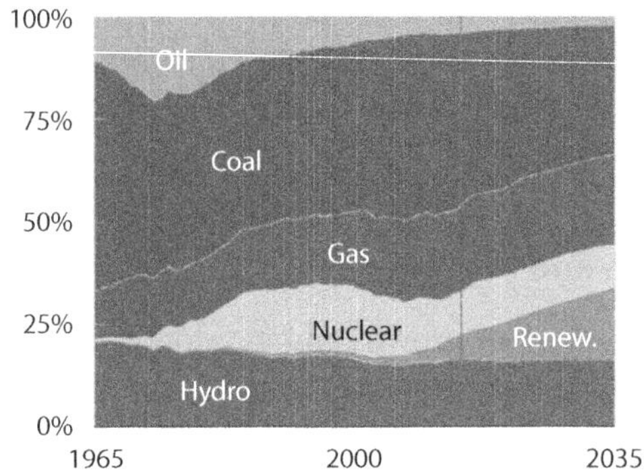

Figure 1.15 Fuel mix in global generation of energy [39].

on the production and utilization of coal ash in the world. Because of rapid industrialization, countries like China and India are showing an increasing demand for coal.

In the Asia Pacific region, fly ash has accounted for a larger share in the global market since 2015 (Figure 1.17). This is particularly due to its use for constructional purposes brought on by a growing population and urbanization. North America, Europe, the Middle East, Africa and the rest of the world are also expected to increase fly ash consumption to fulfill the amount required in construction sectors.

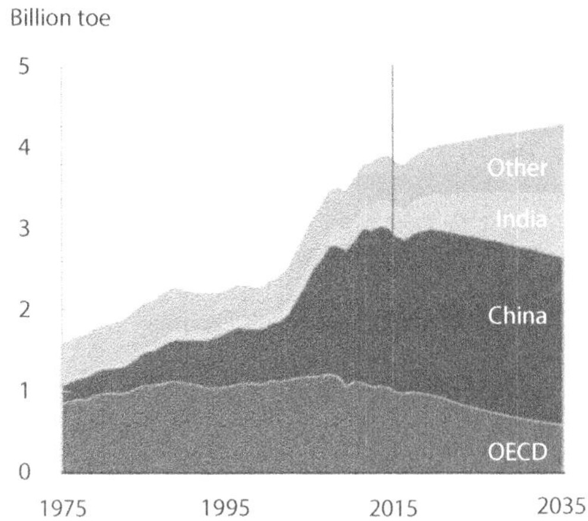

Figure 1.16 Regional breakdown of coal consumption [39].

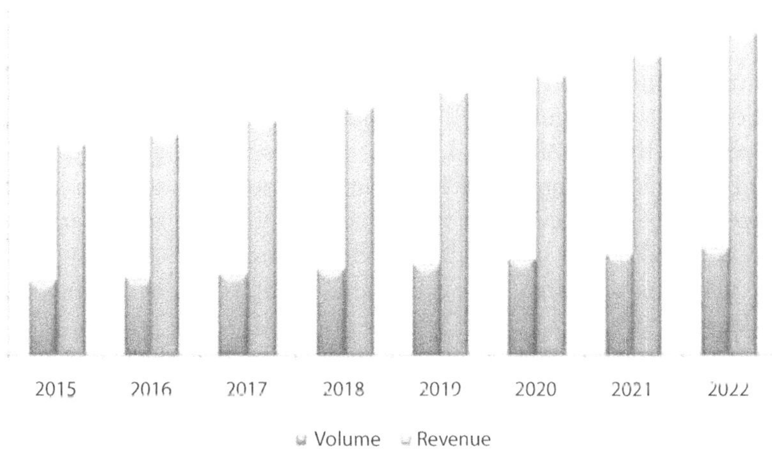

Figure 1.17 Global fly ash market revenue and growth [40].

Coal combustion generates a large volume of solid wastes globally, including fly ash, bottom ash and pond ash. Presently, fly ashes are used in different revenue-producing areas such as road construction, decorative glasses, fire bricks, agriculture sectors, etc. Figure 1.18 shows global fly ash market revenue with areas of application.

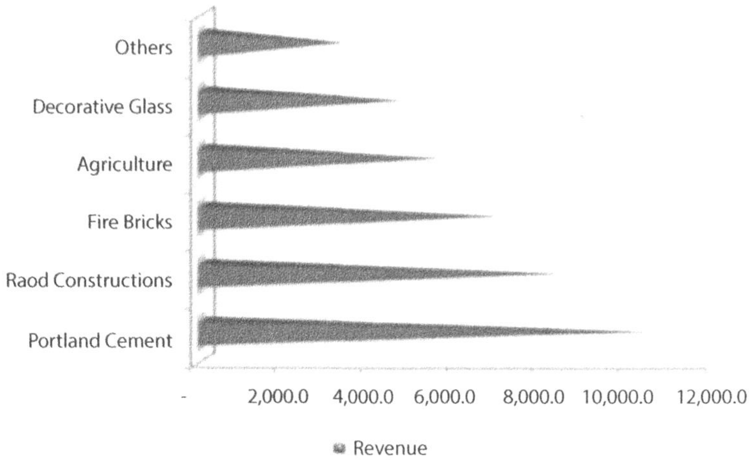

Figure 1.18 Global fly ash market revenue with potential applications [40].

1.3.8 Fly Ash Generation

During thermal power production (Figure 1.19), ashes are produced after coal combustion (Figure 1.20). Typically, coal is fed into the combustion chamber. In the chamber, coal rapidly touches off and generates a warm liquid mineral. The produced liquid mineral solidifies to form slag over different stages, i.e., evaporation and cooling the pipe gas. Coarse particles are called cinder or slag. Lighter fine particles are called fly slag. Fly fiery debris is displaced by electrostatic precipitators or channel fabric baghouses.

Initially, fly ash is released into the atmosphere directly. In recent years, pollution control devices are mandated. The pollution control system captures fly ash before it is released and dumped. Dumped fly ash is a serious environmental issue. Huge acres of land have already been used for the disposal of fly ash and many more acres will be required to accommodate it in the future.

1.3.9 Nature and Composition of Thermal Power Plant Ashes

Fly ash is the fine powder recovered from the gases of coal-fired plants during the production of electricity by electrostatic precipitators. The micron-sized elements of silica, alumina and iron are present in greater amounts; whereas sodium, potassium, titanium, etc., are present in trace amounts. There is a substantial amount of noncombustible impurities present in coal in the form of limestone, shale, dolomite, feldspar and quartz.

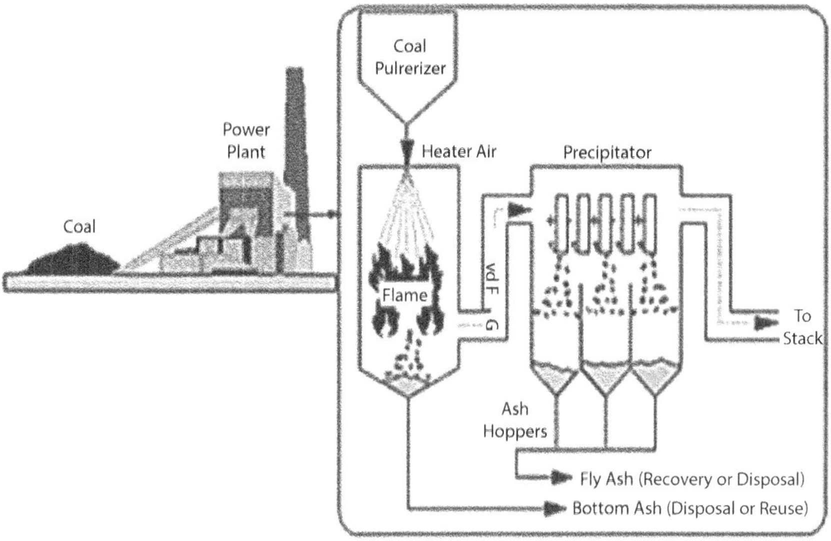

Figure 1.19 Generation of ash at thermal power station [41].

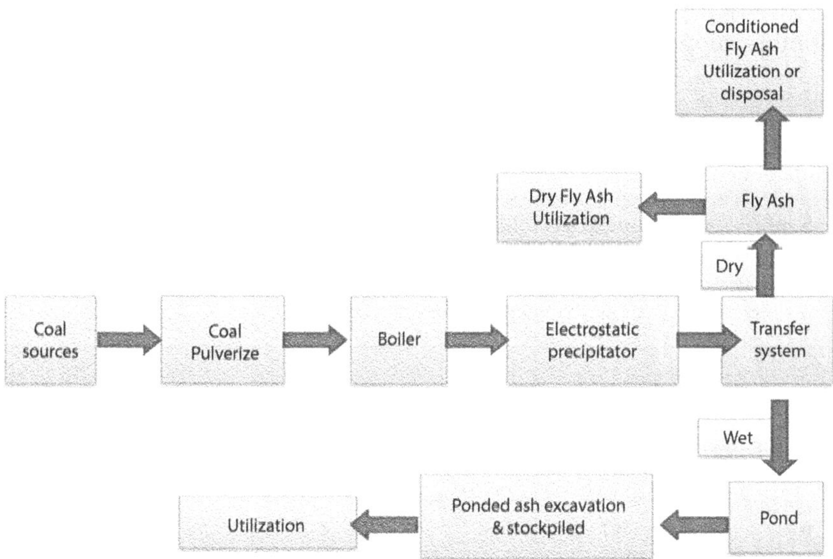

Figure 1.20 Method of fly ash transfer (dry, wet or both conditions) [42].

As the fuel travels through the high-temperature zone in the furnace, the volatile matter and carbon are burnt off, whereas the mineral impurities are carried off in the form of ash by the flue gas. The ash particles become fused in the combustion zone of the furnace; however, upon leaving the combustion zone of the furnace, the molten ash is cooled rapidly and solidifies as spherical glassy particles. Some of the fused matter agglomerates to form bottom ash, but most of it flies out with the flue gas stream and is therefore called fly ash. The fly ash is removed from the flue gas by means of a series of mechanical separators followed by electrostatic precipitators or bag filters. The ratio of fly ash to bottom ash is 72:28 in wet bottom boilers or dry bottom boilers. The chemical composition of fly ash using different types of coal is shown in Table 1.1.

According to ASTM C618, two classes of fly ash are named, *viz.*, Class F fly ash and Class C fly ash. The key difference between these classes is the amount of calcium, silica, alumina, and iron present in ash. The chemical behavior of fly ash depends on which type of coal is burnt (i.e., anthracite, bituminous, and lignite) [43].

To generate Class F fly ash, anthracite and bituminous coal (harder, older) is used. Fly ash (Class F) is pozzolanic in nature. It contains a lower percentage of lime (CaO) (i.e., 7%) [44].

Because of the pozzolanic nature of Class F fly ash, glassy silica and alumina supports the cementing agent. Quick-lime or hydrated lime is

Table 1.1 Normal range of chemical composition of fly ash produced from different coal types.

S. no.	Component	Bituminous	Subbituminous	Lignite
1	SiO_2 (%)	20–60	40–60	15–45
2	Al_2O_3 (%)	5–35	20–30	20–25
3	Fe_2O_3 (%)	10–40	4–10	4–15
4	CaO (%)	1–12	5–30	15–40
5	MgO (%)	0–5	1–6	3–10
6	SO_3 (%)	0–4	0–2	0–10
7	Na_2O (%)	0–4	0–2	0–6
8	K_2O (%)	0–3	0–4	0–4
9	LOI	0–15	0–3	0–5

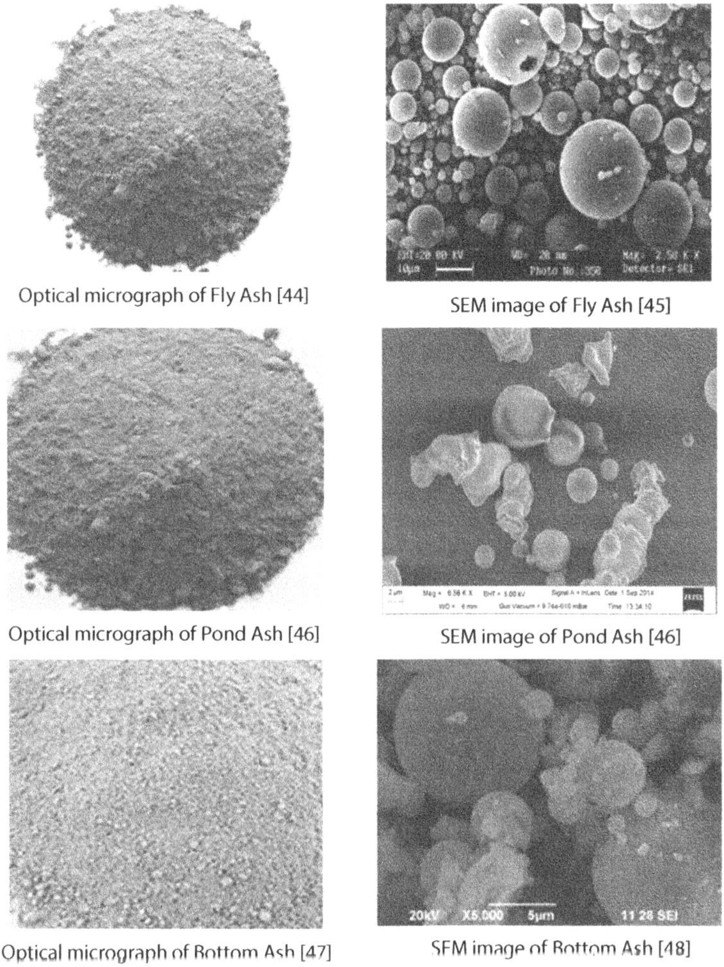

Figure 1.21 Types of thermal power plant ashes (with corresponding SEM images).

mixed with water to form cementitious compounds. Particularly, addition of alkaline activator (i.e., sodium silicate) to Class F fly ash can produce geopolymer.

In Class C fly ash generation, lignite or subbituminous coal (younger) is used to produce thermal power. It has pozzolanic as well as self-cementing properties. If Class C fly ash comes in contact with water, it becomes hardened and stronger over time. In Class C fly ash [44], lime content is more than 20%. Because of its self-cementing behavior, Class C fly ash does not need an activator. In addition, the percentage of alkali and sulfate is higher

Table 1.2 Typical composition of fly ash, different types of fly ash (Class F and Class C), and Portland cement.

Components	Class F fly ash		Class C fly ash		Portland cement	
	Typical*	ASTM C-618	Typical**	ASTM C-618	Typical***	ASTM C-150
SiO_2	48.0	---	37.3	---	20.25	---
Al_2O_3	24.3	---	21.4	---	4.25	---
Fe_2O_3	15.6	---	5.7	---	2.59	---
$SiO_2 + Al_2O_3 + Fe_2O_3$	87.9	70.0 (min.%)	64.3	50.0 (min.%)	---	---
CaO (Lime)	3.2	---	22.4	---	60.6	---
MgO	---	---	---	---	2.24	6.0 (max.%)
SO_3	0.4	5.0 (max.%)	2.5	5.0 (max.%)	---	3.0 (max.%)
Loss of Ignition (LOI)	3.0	6.2 (max.%)	0.4	6.0 (max.%)	0.55	3.0 (max.%)
Moisture content	0.1	3.0	0.1	3.0	---	---
Insoluble residue	---	---	---	---	---	0.75
Available alkalis as equivalent Na_2O	0.8	1.5	1.4	1.5	0.20	---

than Class F fly ash. Typical compositions of different types of fly ash (Class F and Class C) and Portland cement are indicated in Table 1.2.

Some quantity of fly ash utilization in globally has now picked up. Still, there special attention is needed for 100% utilization of ash.

Bottom ash is another type of ash produced during power generation. It is a noncombustible residue obtained from coal combustion. It comprises traces of combustibles embedded in forming clinkers and sticking to hot side walls of a coal-burning furnace during its operation. However, the portion of bottom ash that escapes up the chimney is referred to as fly ash. Clinkers fall into the bottom hopper of a coal-burning furnace and are cooled. This ash is also termed as bottom ash.

Clinker lumps are crushed in a water impounded hopper (WIH) system and crushed into small sizes by clinker grinders mounted under water, and drop into a trough. From there a water ejector pumps them out to an ash pond used as a landfill. In another arrangement a continuous link chain scrapes out the clinkers from under water and feeds them into a clinker grinder outside the bottom ash hopper.

Modern systems adopt a heavy-duty chain conveyor submerged in a water bath (below the furnace). It quenches hot ash as it falls from the combustion chamber and removes wet ash continuously by dewatering slope before discharging it onto mechanical conveyors or directly to storage silos.

1.3.10 Pond Ash

Pond ash [46] is another waste product obtained from boilers. It is the wet fly ash mixed with bottom ash that is disposed of in large ponds or dykes as slurry. In the present scenario, pond ash is being generated at an alarming rate. Generated pond ash poses a great threat to the environment. Thus, its sustainable management has become a key focus area in materials engineering research. As pond ash is relatively coarse and contains dissolvable alkalis, it is washed with water; hence, it has low pozzolanic reactivity. Due to its low pozzolanic effect, it is chosen as a partial replacement of cement to make concrete.

There is the possibility of using pond ash in different areas. One area is for producing burnt clay bricks. During the production of burnt clay brick, a green brick is prepared by appropriately mixing clays and pond ash and is made by a conventional method (at a brick manufacturing plant). Pond ash is also called coal ash [49].

Another application involves controlling air pollution and the pozzolanic effect of the ashes. In this application, a pond is used as a landfill to capture the ash [50]. Due to gravity. ashes settle in ponds. This technology does not treat dissolved pollutants [51]. Usually, ponds have not been built as lined landfills. Hence, chemicals in the ash can leach into groundwater or surface waters, accumulating in the biomass of the system [52–54].

Ash ponds are generally formed using a ring embankment to enclose the disposal site. Embankments are designed using similar design parameters as embankment dams. The design process is primarily focused on handling seepage and ensuring slope stability. Failure of a pond's earthen embankment can cause ash spills on adjacent land and rivers, with serious environmental damage, as evidenced in the 2008 Kingston Fossil Plant coal fly ash slurry spill in Tennessee [55] and the 2014 Dan River coal ash spill in North Carolina [56].

In fact, pond ash is a mixture of fly ash and bottom ash. Types of thermal power plant ashes (with corresponding SEM images) is shown in

Table 1.3 Chemical composition of the NALCO ash pond [46].

Raw Materials	SiO_2	Al_2O_3	CaO	MgO	Fe_2O_3	TiO_2	Cr_2O_3	MnO	P_2O_5	C	LOI
Pond Ash	62.8	28.3	0.7	0.58	3.85	1.84	0.04	0.03	0.32	1.15	0.5

Table 1.4 Chemical and physical characteristics of pond ash.

Chemical characteristics [46]		**Physical characteristics [57]**	
Parameters	**Concentration (wt%)**	**Parameters**	**Pond ash**
SiO_2	59.007		
Al_2O_3	19.551	Specific gravity @ 27°C	2.0675
Fe_2O_3	15.350	Fineness (m^2/kg)	73.78
TiO_2	3.158	Hydraulic conductivity @ 27°C	0.992
K_2O	1.271	Dry density, γd (g/cc)	0.848
CaO	1.151	Void ratio	1.435
Mn_2O_3	0.197		

Figure 1.21. The main difference between pond ash and fly ash is particle size. Pond ash is coarser and has less of a pozzolanic effect.

The process of generating pond ash, such as facility, variation in collection, disposal and storage methods, temperature of coal burning, and peak load demand in thermal stations, etc., depends on the type and source of coal used during the production of thermal power. Engineering properties of pond ash control its use in the area of construction. To evaluate engineering performance, the physical, chemical, mineralogical and morphological properties of pond ash play an important role in assessing its suitability for applications. The chemical composition of pond ash is presented in Table 1.3. Also, the chemical and physical characteristics of pond ash are indicated in Table 1.4.

1.3.11 Various Uses of Pulverized Fuel Ash

Pulverized fuel ash is utilized in diverse sectors. Pozzolanic behavior of fly ash makes it a resource material for preparing cement and other fly ash-based products. Also, bottom ash, pond ash and coarse fly ash allow its use in construction of embankments, structural fills, reinforced fills, low-lying area development, etc., because of its geotechnical properties. The physicochemical properties of pond ash are similar to soil. Pond ash contains P, K, Ca, Mg, Cu, Zn, Mo, and Fe, etc., which are essential nutrients for plant growth. Therefore, it is used as a soil amender and source of micronutrients in agriculture/soil amendment.

Major application areas of pond fly ash (PFA) or pond ash are as follows:

1) Manufacture of Portland Pozzolana cement and performance improver in Ordinary Portland Cement (OPC);
2) Partial replacement of OPC in cement concrete;
3) High-volume fly ash concrete;
4) Roller-compacted concrete used for dam and pavement construction;
5) Manufacture of ash bricks and other building products;
6) Construction of road embankments, structural fills, low-lying area development;
7) As a soil amender in agriculture and wasteland development.

Particularly, pond ash and residue of an integrated steel plant are mixed for manufacture of bricks, making dry lean concrete as base course in four/six lane highways.

Pond ash is potentially useable material. The variable characteristics of pond ash are obtained from disposal, differences in the unit weight of fly ash and bottom ash, ash particle size, setting time and suspension time. In addition, the type and source of coal, performance of generating facility, variation in collection, disposal/storage methods, temperature of coal burning, peak load demand in thermal stations and a few other issues control the properties of pond ash. The engineering properties of pond ash control its use as a material in construction. Hence, characterization of pond ash in terms of its physical, chemical, mineralogical and morphological properties plays a very important role in assessing its suitability as a material in different fields of construction. Some of the reports available on pond ash are mentioned below.

Langton *et al.* [58] investigated the use of three sources of pond ash in flowable slurry along with cement, sand and water. A cement content of 1% to 12% by weight of fly ash is used in order to develop 0.2 to 1 MPa compressive strength at 28-days setting. Furthermore, Naik and Kraus [59] developed flowable slurry using pond ash, fine crushed sand, cement and water. Mixtures are developed with different combinations of pond ash and fine crushed sand, i.e., 0% coal ash and 100% fine sand, 100% coal ash and 0% fine sand. It is found that water requirement of the mixes decreased with increased fine crushed sand content. Density of the flowable slurry increased as the amount of fine crushed sand increased. Langton *et al.* [60] (1998) developed stetting/shrinkage time of mixture. Pond ash set in 26 hours to 30 hours and shrinkage/settlement is only 5% after 28 days.

Pond ash is not uniform in color, size and texture and is recommended for common backfill; whereas, fine-grained material that is more homogeneous is recommended for both structural backfill and common backfill.

1.3.12 Importance of Pond Ash Management

There is a rising demand for energy in India, which is required to expand the power sector. The power sector requires a huge amount of coal for generating power. A quantum of the pond ash produced in thermal power plants can cause three environmental risks, i.e., air, surface water and groundwater pollution. Air pollution is caused by direct emissions of toxic gases from the power plants as well as windblown ash dust from ash ponds. The airborne dust can settle on surface water or soil and may contaminate the water/soil system. The wet system of disposal in most power plants discharges particulate of ash directly into the nearby surface water system. This has warranted research and development work in both the scientific and industrial communities in order to find innovative uses and safe disposal of pond ash.

1.4 Physical Characteristics of Coal Ashes

The specific gravity of coal ash is one of its key physical properties. Such analysis is needed before coal ash is used in geotechnical and other applications [61]. Typically, the specific gravity of coal ash is around 2.0. but varies to a large extent from 1.6 to 3.1 depending on a combination of factors such as gradation, particle shape and chemical composition. It mostly consists of glassy cenospheres and some solid spheres. The reason for the low specific gravity of coal ash might either be the presence of the large number of hollow cenospheres (entrapped air) or variation in its chemical composition (particularly iron content). Generally, the specific gravity of fly ash is higher than pond ash and bottom ash if collected from the same place. When the particles are crushed, they show a higher specific gravity compared to the uncrushed portion of the same material [62].

The morphology of coal ash is particle in nature. The form or structure of particle morphology is helpful to understand the physical properties and leaching behavior of coal ash [63].

The surface area of coal ash particles is important for controlling total adsorption capacity. The surface area of coal ash particles is inversely related to the particle size; the smaller the particle size, the larger the surface area [64].

The permeability or hydraulic conductivity (K) is often used to study the interchangeable property of coal ash. Coefficient of permeability or hydraulic conductivity (K) takes into account the fluid properties. Intrinsic permeability (k) only denotes the effectiveness of the porous medium alone [65]. Coefficient of permeability and intrinsic permeability are related as follows;

$$K = k \times \text{specific weight of fluid/dynamic viscosity of fluid}$$

Fluid actually flows through void spaces. So, porosity can play a key role in controlling the influence of permeability. Porosity value of a sample is representative of water bearing capacity. Porosity is generally designated by 'n'. It is the ratio between volumes of voids and total volume (multiply by 100 to express as percent). Evidently, porosity and permeability are affected by compaction (density), since compaction reduces the amount of void or space for a given total volume. Normally, the higher porosity samples will have higher hydraulic conductivity. Fly ash compacted in a laboratory to 95% maximum density can achieve a permeability of 1×10^{-5} cm/sec. Higher density results in a lower permeability, since low permeability will restrict leachate from migrating away from the site. Similar results are found in compacted fly ash, which generally has a low hydraulic conductivity at 9 cm/day (1×10^{4} cm/sec), whereas uncompacted fly ash can be found as 70 cm/day (8.1×10^{4} cm/sec).

Fly ash is used (to change soil) for plant growth. Hydraulic conductivity in soil increased up to 10 to 20% fly ash by volume than decreased as a result of fly ash's pozzolanic reaction, which tends to cement when in contact with water. A small amount of fly ash is needed in acidic soils for this purpose. Low hydraulic conductivity is desirable if fly ash is to be reused. Even in a disposal situation, low hydraulic conductivity would discourage water from seeping through and forming leachate. A higher hydraulic conductivity layer surrounding the fly ash lowers the resistance.

- *Density* is also affected by compaction. In a bituminous coal-based fly ash, 95% maximum density (1.3 g/cm^3) is achieved [61]. In a disposed of state, 85% maximum density (1.1 g/cm^3) is achieved. Since fly ash generally has a low bulk density, when added to soil it reduces the bulk density of soil.
- *Grain size distribution* [66] indicates whether coal ash is well graded, poorly graded, fine or coarse, etc., which helps to classify coal ashes. Coal ash is predominantly silt-sized with some sand-sized fraction. Leonards and Bailey [67] have reported the range of gradation of fly ash and bottom ashes, also known as silty sands or sandy silts. An extensive investigation was carried out on Indian coal ashes. Fly ash particles consist of a silt-size fraction with some clay-size fraction. Both fly ash and pond ash particles consist of a silt-size fraction with some sand-size fraction, whereas bottom ash is coarser particles and consists of a sand-size fraction with some silt-size fraction. Based on grain-size distribution, coal ash is classified as sandy silt to silty sand. It is poorly graded with coefficient of curvature ranging between 0.61 and 3.70; and coefficient of uniformity in the range of 1.59–14.0.
- *Free swell index* [68] in soil engineering serves as a tool to identify swelling soils. Free swell test was proposed by Holtz and Gibbs to estimate swell potential. Sridharan and K. Prakash [69] modified the definition of free swell index,

with interesting results; 70% of coal ash shows negative free swell index due to flocculation. This is because the clay-size fraction in coal ash is much less and the free swell index is negligible.

- *Index properties* [61] are extensively used in geotechnical engineering practice. Liquid limit is one of the main physical properties and is used to correlate the engineering properties. Percussion cup and fall cone methods are popular to determine the liquid limit of fine-grained soils. The percussion cup method is not suitable for studying fly ash due to the non-plastic nature of fly ash. The cone penetration method mainly used to study fly ash is not suitable to determine index properties. Since it is very difficult to get a smooth level surface, a new method is used to determine the liquid limit, which is called the "equilibrium water content under K_0 stress method." It has been found to be an effective method for determining liquid limit of coal ashes. The method is simple, reasonably error free, less time-consuming and has good reproducibility (except for Class C fly ashes). Using this method, liquid limit water content results of coal ashes are found to be 26–51% (for fly ashes), 22–64% (for pond ashes), and 45–104% (for bottom ashes). Liquid limit values showed by coal ashes and are due to their fabric and carbon content. The tested coal ashes are non-plastic, and hence plastic limit could not be determined.

  Also, it is not possible to carry out shrinkage limit tests. During drying, ash pats are crumbled. So, amount of shrinkage is much less and shrinkage limit will be quite high. Hence, shrinkage will not be a constraint.
- *Specific surfaces* [61] of coal ash are primarily silt/sand-sized particles and their specific surface is expected to be very low. Therefore, surface area measurements need to be obtained for completeness and use in certain cases. Surface area measurements are done using two methods, i.e., Desiccator method and Blaine air-permeability method.
- *Chemical properties* of coal ashes significantly influence their environmental impacts. Therefore, the need has arisen to investigate the engineering properties of coal ash in relation to its use and disposal. Its adverse impacts include contamination of surface and subsurface water with toxic heavy metals present in coal ashes, loss of soil fertility around the

plant sites, etc. Hence, studies on chemical composition, morphology, pH, total soluble solids, etc., are essential [61].

- *Chemical composition* suggests possible applications of coal ash. All the Indian coal ashes need to satisfy chemical requirements before their use [61].
- *Mineralogical phases* are primarily identified by X-ray diffraction studies. Coal ashes predominantly contain quartz and feldspar minerals [70] and show crystalline as well as amorphous phases. Fly ash has a particularly amorphous ferro-aluminosilicate mineral composition. Amorphous iron aluminum oxides and manganese oxides present on fly ash surface act as a sink. Trace element is available for leaching. Degree of solubility of these oxide sinks determines the elements released into the aqueous medium.

  The pH [71] of the aqueous medium affects solubility of these oxides, i.e., physicochemical characteristics. Furthermore, mobilization of trace elements in aqueous medium is regulated by solubility of hydroxide and carbonate salts. It also depends on the pH of aqueous media. Fly ash has higher pH value compared to pond ash and bottom ash. This is due to the presence of higher amounts of free lime and alkaline oxides. Since all types of coal ash are nearly alkaline it is used in reinforced cement concrete to protect against corrosion.
- *Solubility of solids* greatly influences engineering properties. Also, solubility of nutrient elements such as calcium, magnesium, iron, sulfur, phosphorus, potassium and manganese affects crop yield to a great extent. Solubility of solids is found to be 400–17600 ppm for fly ash, 800–3600 ppm for pond ash, and 1400–4100 ppm for bottom ash.
- *Strength of fly ash* improves with time due to pozzolanic reactions [72]. Reactive silica and free lime contents are necessary for pozzolanic reactions to take place. Lime reactivity is a property which depends on the proportion of silica reactive in coal ash. The pozzolanic effect is higher in fly ash but not bottom ash and pond ash. The higher percentage of free lime in coal ash plays an important role in the reactivity.
- *Compaction behavior* [73] of coal ash is an important parameter to control the strength, compressibility and permeability properties. Densification of ash improves the engineering properties. Compaction of material greatly affects the energy

application, grain size distribution, plasticity characteristics and moisture content. Variation of dry density with moisture content of fly ash is less than that of well-graded soil, which could be explained by the presence of higher air void in fly ash. Normally, soils have air voids. Dry density of soils ranges from 1–5%, whereas dry density of fly ash ranges from 5–15%. Since higher void content tends to limit buildup of pore pressures during compaction, fly ash should be compacted over a larger range of water content.

- *Strength behavior* [74] is also an important engineering property. Compressive strength of fine ash is higher than that of coarser ash-based materials. Also, most of the shear strength of coal ash is due to internal friction. Effects of additives, such as gypsum and lime on ash-based materials, have been shown to increase strength. In some cases, addition of gypsum on ash-based specimens has no effects on the strength characteristics.
- *Permeability* [75] is an important parameter in the design of liners. It contains leachate migration, dykes (to predict the loss of water and stability of slopes) and sub-base material. Coefficient of permeability of ash depends on grain size, degree of compaction and pozzolanic activity. Permeability is found to be 8×10^{-6} cm/s to 1.87×10^{-4} cm/s (for fly ash), 5×10^{-5} cm/s to 9.62×10^{-4} cm/s (for pond ash), and 9.9×10^{-5} cm/s to 7×10^{-4} cm/s (for bottom ash).
- *Leaching behavior* means water passes through a porous media. In other ways, each constituent present in the matrix dissolves into pore water at some finite rate because there is no such thing as a completely insoluble material [76]. Permeation of contaminated pore water of porous matrix due to any driving force is called "leaching." Any contaminated water generated that passes through a porous matrix is called "leachate." The capacity of waste material to leach is called its "leachability." Depending on sources of coals used in thermal power plants, fly ash may contain different toxic elements. Leaching of these toxic elements from ash to pond(s) is of considerable importance because it may cause serious environmental problems. Due to several factors, such as quality of coal, sources of water, pH, time, temperature, etc., the release of these contaminants and their subsequent influence affect groundwater quality. Leachate

characteristics are highly variable, and even within a given landfill site, leachate quality varies over time and space [76].

1.5 Coal Ash Utilization

Coal ash uses include (in approximate decreasing order of importance) [77]:

- Concrete production, as a substitute material for Portland cement and sand
- Embankments and other structural fills (usually for road construction)
- Grout and flowable fill production
- Waste stabilization and solidification
- Cement clinkers production (as a substitute material for clay)
- Mine reclamation
- Stabilization of soft soils
- Road sub-base construction
- As aggregate substitute material (e.g., for brick production)
- Mineral filler in asphaltic concrete
- Agricultural uses in soil amendment, fertilizer, cattle feeders, soil stabilization in stock feed yards, and agricultural stakes
- Loose application on rivers to melt ice
- Loose application on roads and parking lots for ice control

Other applications include cosmetics, toothpaste, kitchen countertops, floor and ceiling tiles, bowling balls, flotation devices, stucco, utensils, tool handles, picture frames, auto bodies and boat hulls, cellular concrete, geopolymers, roofing tiles, roofing granules, decking, fireplace mantles, cinder block, PVC pipe, structural insulated panels, house siding and trim, running tracks, blasting grit, recycled plastic lumber, utility poles and cross arms, railway sleepers, highway sound barriers, marine pilings, doors, window frames, scaffolding, sign posts, crypts, columns, railroad ties, vinyl flooring, paving stones, shower stalls, garage doors, park benches, landscape timbers, planters, pallet blocks, molding, mail boxes, artificial reef, binding agent, paints and under coatings, metal castings, and filler in wood and plastic products.

More recently, fly ash has been used as a component in geopolymers, where the reactivity of the fly ash glasses generates a binder. It possibly reduces CO_2 emissions.

1.6 Slag

Slag is a glass-like by-product left over after separation of a desired metal from its ore. Slag is usually a mixture of metal oxides and also can contain metal sulfides and elemental metals. Slag is usually used to remove waste in metal smelting, and can also assist in controlling smelting temperature and minimizing any reoxidation of the final liquid metal product. Slag is used to make solid metal. In some smelting processes, such as ilmenite smelting, titanium dioxide is produced. The slag is a more valuable product than the metal [78].

In nature, metals like iron, copper, lead and nickel are found in impure states, called ores. During smelting, ore is exposed to high temperatures, and impurities are separated from the molten metal. Slag is the collection of compounds that are removed as by-product. In the smelting processes, oxides are introduced to control the slag chemistry. In many smelting processes, oxides are introduced to control the slag chemistry, assisting in the removal of impurities and protecting the furnace refractory lining from excessive wear. In this case, the slag is called synthetic. In steelmaking, slag, quick lime and magnesite are introduced for refractory protection. In neutralizing the alumina and silica, it is separated from the metal and helps to remove sulfur and phosphorus from the steel.

In ferrous and non-ferrous smelting processes, slags are produced. Smelting of copper, lead and bauxite (non-ferrous process) removes iron and silica and separates them as iron-silicate-based slags. Slag from steel mills (in ferrous smelting) is designed to minimize iron loss and mainly contains oxides of calcium, silicon, magnesium, and aluminum. The quartz component of the original ore automatically goes through the smelting process as silicon dioxide.

As the slag is channeled out of the furnace, water is poured over it. This rapid cooling occurs at 2,600 °F (1,430 °C). It causes several chemical reactions to take place within the slag. It gives material which has cementitious properties.

Water carries slag in its slurry form to a large agitation tank. It is then pumped along a piping system into a number of gravel-based filter beds. Such filter beds then retain slag granules. Water drains away from it and is returned to the system.

1.6.1 Generation of Slag

In India, the steel industry produces about 24 million tonnes of blast furnace (BF) slag and 12 million tonnes of steel slag annually. It is believed that generated BF slag may reach around 45–0 million tons by 2030, whereas basic oxygen furnace (BOF) slag is predicted to be around 15–20 million tons per year by the same time. Compared to the present level of around 5 million tons per year, generation of electric arc furnace (EAF) slag and iron slag also will increase to more than 10 million tons per year by 2030. In the past few years, steel slag has been utilized in very low amounts. In particular, sintering has resulted in the accumulation of large quantities of steel slags in every plant. It is estimated that more than 30–40 million tons of steel slag may be lying around in various steel plants. Slag is required

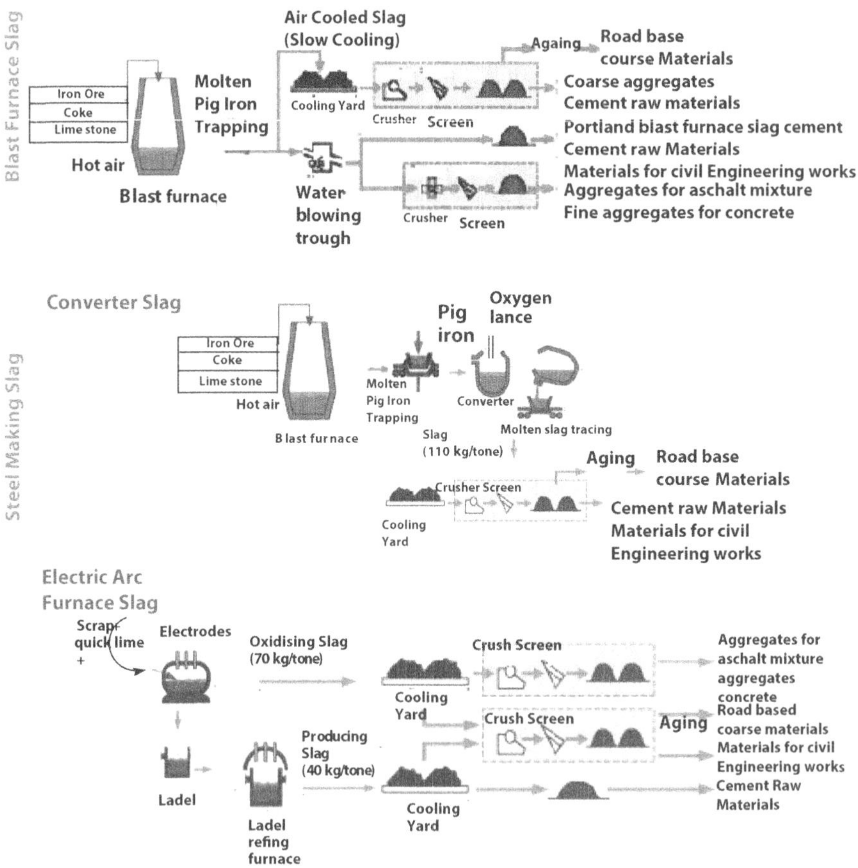

Figure 1.22 Typical slag generation and utilization [79].

in the cement industry, for making roads, and in the agricultural sector. Typical slag generation and its utilization across the globe are shown in Figure 1.22 [79].

In addition to slag, dust and sludge is also generated in various processes. Slag amount varies from place to place and depends on the adoption of technology. Average generation of slag, sludge and dust from various processes is shown in Figure 1.23. Most steel industries are taking steps to control the generation of slag; and even though significant progress has been made, the majority of units are yet to achieve benchmark performance.

- *Blast Furnace Slag (BFS)* [85] is produced during the making of iron in a blast furnace. When it is removed from the blast furnace, this slag is molten and is at a temperature of

Optical photo of air-cooled slag [80].

Optical photo of granulatedslag [80].

Optical photo of steelmaking slag [81].

Optical photo of electric furnace slag [82].

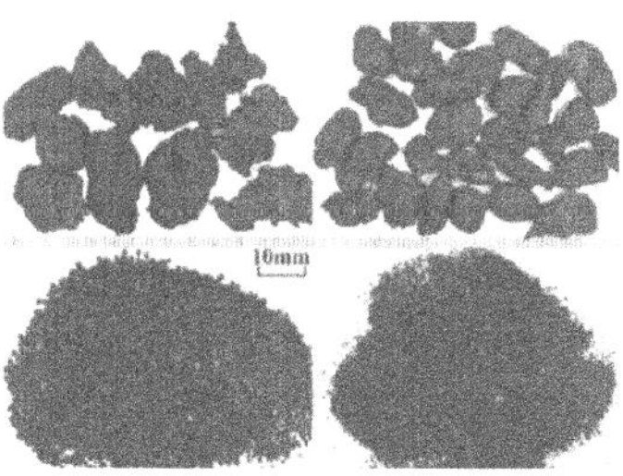

Optical photo of converter slag [83].

Optical photo of steelmaking slag [84].

Figure 1.23 Optical photo images of various types of slag.

approximately 1500 °C. Blast furnace slag can be cooled in different ways to form any of several types of BF slag products, as follows:

- In Air-Cooled BF Slag (ACBFS) [86], molten BFS is allowed to flow from blast furnace to open air pits located beside the furnaces, where the material is quenched with water applied by sprays to facilitate cooling. Alternatively, molten slag is dumped in an open yard where it is allowed to cool naturally. Once sufficiently cooled, it results in a crystalline, rock-like air-cooled slag and is termed rock slag. The ACBFS is dug and transported to a nearby crushing and screening (aggregate) plant, where it is processed into aggregates. However, open air cooling requires large areas of land. On the other hand, land availability is getting scarcer day by day and the pollution control board has also been impressing upon the steel sector to avoid dumping of slag in open land.
- Physical properties of ACBFS [85] are similar to other natural rock and is used as coarse and fine aggregate. Hence, it is used as substitute of natural aggregates in all types of road constructions. Granulated ground BF slag is also used as substitute to natural sand for use in civil works. In addition, BF slag is also used in embankment, land fill, etc.

- *In Granulated Blast Furnace Slag* (GBS or GBFS) [87], molten slag (modern blast furnaces) is directly converted into fine granules in BF shop complex. Granulated BFS (GBFS) can be used for manufacture of slag cement. Ground granulated blast furnace slag (GGBFS) or fine aggregates are employed for making concrete.
- *In Linz-Donawitz (LD) or Steel Slag* [88], basic oxygen steelmaking (BOF) slag is also called Linz-Donawitz slag. It is cooled in a cooling yard through air cooling and water sprinkling (moderate). However, this work requires a significant amount of time to do the cooling. To cool hot slag down to a workable temperature demands the allocation of a spacious land/yard. It is also discouraged by company/industry authorities. For their operational safety, a number of more efficient cooling processes have been required.

Table 1.5 Iron and steel slag compositions [94].

Type of components	Blast furnace slag (BFS)	Converter slag (CS)	Electric arc furnace slag		Andesite (for reference)	Ordinary cement
			Oxidizing slag	Reducing slag		
CaO	41.7	45.8	22.8	55.1	5.8	64.2
SiO_2	33.8	11.0	12.1	18.8	59.6	22.0
T-Fe	0.4	17.4	29.5	0.3	3.1	3.0
MgO	7.4	6.5	4.8	7.3	2.8	1.5
Al_2O_3	13.4	1.9	6.8	16.5	17.3	5.5
S	0.8	0.06	0.2	0.4	---	2.0
P_2O_5	<0.1	1.7	0.3	0.1	---	---
MnO	0.3	5.3	7.9	1.0	0.2	---

- *Iron and Steel Slag* [89, 90] is generated as a by-product during the manufacturing of iron and steel. The slag can be categorized into blast furnace slag and steelmaking slag. Originally, slag was produced by metal manufacturing processes. However, it now originates from molten waste material when trash. Other substances are disposed of at an incinerator facility. Iron and steel slag compositions are presented in Table 1.5.
- *In Converter Slag* [91, 92], the slag is cooled slowly by natural cooling by spraying water in a cooling yard. It is then processed and used in different iron and steel slag applications. Approximately 110 kg of slag is produced during the production of one ton of converter steel.
- *Electric Arc Furnace Slag* [93] is generated when iron scrap is melted and refined. It consists of oxidizing slag which is generated during oxidation refining. Reducing slag is produced during reduction refining. Approximately 70 kg of electric arc furnace oxidizing slag and 40 kg of reducing slag are generated for one ton of electric arc furnace steel.

1.6.2 Slag Properties and Utilization

Composition of steel slag varies with furnace type, steel grades and pretreatment method. However, the main constituents remain SiO_2, CaO, Fe_2O_3, FeO, Al_2O_3, MgO, MnO and P_2O_5 [94]. In steel slag, the most important mineral phases are dicalcium silicate (Ca_2SiO_4), tricalcium silicate (Ca_3SiO_4), RO phase (CaO-FeO-MnO-MgO solid solution), tetra-calcium aluminoferrite (Ca_4AF), olivine (fayalite (Fe_2SiO_4); few have kirschsteinite ($CaFeSiO_4$) compositions, merwinite and free lime [95].

Granulated blast furnace slag is extensively used in manufacture of slag cement (up to 70% addition is permitted as per IS-455). Granulated blast furnace slag is a latent hydraulic material and is glassy in character. Its glass content varies from 90–95%. Blast furnace slag is fully utilized in cement. Other applications of steel slags are road construction, sand, insulation wool, and agriculture. LD/BOF slag is used in road projects and iron-making via recycling. Steelmaking slag is used for steel slag as aggregate in civil construction. Steel slag is also traditionally used in road construction, in rail ballast, as soil stabilizer (silica and phosphorous supplement), as performance improver in cement making, and as raw material in clinker manufacture replacing lime stones. Steel slag is being explored widely across the globe, mainly for supplementing soil nutrients as well as soil remediation of acidic soil. Steel LD slag can also be used in fertilizers for agricultural applications.

1.7 Geopolymers

The term "geopolymer" was coined by the French scientist Joseph Davidovits in 1978 [96]. Geopolymers are alumino-silicate or inorganic polymers with an amorphous microstructure. They are prepared in an alkaline environment. In an extensive study, Davidovits developed the chemistry of geopolymer science. The properties of geopolymers were established in 1979 as new generation material. The term poly(sialate) is the chemical designation of geopolymers, i.e., silico-aluminates [97]. Sialate is an abbreviation for alkali silicon-oxoaluminate in which the alkali is sodium-potassium-lithium-calcium [97]. Polysialates are chain and ring polymers with Si^{4+} and Al^{3+} in IV-fold coordination with oxygen. Polysialate has the empirical formula: Mn $(-(SiO_2)z-AlO_2)_n \cdot wH_2O$.

Amorphous to semicrystalline three-dimensional silico-aluminate structures are known as geopolymers [98]. Furthermore, the chemical composition of geopolymer is similar to zeolite [99].

Usually, geopolymers harden via polymerization, but not in the hydration process (as OPC-based binders). Most researchers were uncertain of the reaction mechanism and kinetics governing the geopolymerization process. However, back in the 1950s, Glukhovsky [100] proposed an alkali activation mechanism for geopolymer materials with silica and reactive alumina as the primary components.

In the model, the geopolymerization process is divided into three phases:

- Destruction–coagulation
- Coagulation–condensation
- Condensation–crystallization

A simplified reaction mechanism of geopolymerization has been proposed [101–103]. This conceptual model is shown in Figure 1.24. Different steps lead to the polymerization and hardening of the aggregates to concrete. The geopolymerization process requires the chemical reaction of alumino-silicate oxides (Si_2O_5, Al_2O_3) with alkali. Polysilicate leads to polymeric Si-O-Al bonds. There are further examples of polycondensation by alkali into polysialate-siloxo [101–103]. Nonetheless, the most popular conceptual model proposed for setting and hardening of geopolymer materials consists of the following phases [102, 103]:

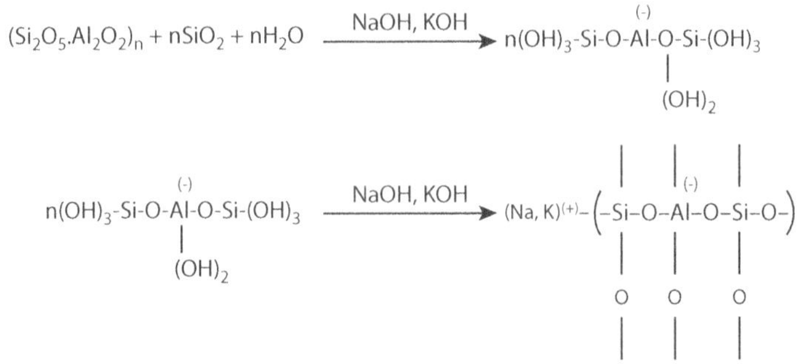

Figure 1.24 Schematic representation of geopolymer reaction [104].

- Dissolution of Si and Al atoms from source material by the action of hydroxide ions.
- Condensation of precursor ions into monomers.
- Polycondensation/polymerization of monomers into polymeric structures.

Water is removed during the geopolymerization reaction that occurs in the formation of geopolymers. This water is expelled from geopolymer matrixes during curing and further drying periods, which leaves behind discontinuous nanopores in the matrix and benefits the performance of geopolymers. Water in a geopolymer mixture plays no role in the chemical reaction and merely provides workability to the mixture during handling.

1.7.1 Constituents of Geopolymers

Ordinary Portland cements (OPC) are consistently used as one of the building raw materials worldwide. Annually, global cement production is 2.8 million tons. It is believed that the production rate is increasing by 4 million tons annually [105, 106]. Available reports indicate that there may be a need for more production of OPC. Therefore, the production rate will sharply rise to 5.5 gigatons per year [105, 107], which will cause major environmental pollution [105–108]. It is reported that 0.54 tons of CO_2 per ton (from cement production industry) and 0.46 tons of CO_2 (from other sources) are liberated. Therefore, cement production alone can contribute approximately 7% CO_2 to our environment [105, 108–112].

Problems brought on by global warming and the greenhouse gas effect have led to several investigations into developing novel constructional binders other than the cement, which will reduce CO_2 emissions [113].

Novel cementitious materials, such as geopolymers, were first developed by Davidovits in 1978 [96]. These materials are characterized by molecular chains and network similar to organic polymers [96]. They are formed by the action of alkali molecules on inorganic compounds bearing silicon and aluminum atoms. The strength of these materials (i.e., GP) is due to chain formation as well as network formation (crosslinking like thermosetting polymer). These materials are found suitable to replace conventional ordinary Portland cement (OPC) [102, 103]. Since these materials can be produced from industrial waste, such as fly ash/pond ash/slag from steel plants, the products will be very cost-effective and will also help to reduce pollution. As such manufacturing of cements is an energy-intensive process, the environment is also affected by the dust it releases into the atmosphere, which is injurious to human health. Thus, geopolymer (GP) can be reformed as alkali-activated cement or inorganic polymer cement. However, the development of these materials is in its infancy. These inorganic products consist of the $[SiO_4]^{-4}$ and $[AlO_4]^{-4}$ tetrahedral networks. The influences of alumino-silicate materials, such as metakaolin and low-calcium fly ash, were investigated by many researchers [114]. Their properties are suitable for reformation of aggregate similar to OPC. Based on the past results, high-calcium fly ash has been successfully used as a raw material in the geopolymer mixture [115]. A wide range of applications of geopolymers [116–122] in the fields of new ceramics, binders, and matrices for hazardous waste stabilization, requiring fire-resistant materials and high-tech materials are an emerging trend of current research.

The increased use of cement in concrete poses an environmental threat due to the emission of carbon dioxide during the manufacturing process. About 1.5 tons of raw materials are needed in the production of every ton of Portland cement (PC); at the same time, about one ton of carbon dioxide (CO_2) is released into the environment during the production [105–108]. The production of PC is a costly and energy-intensive process. Additionally, global warming also occurs because of greenhouse gases, such as carbon dioxide, which accumulate in the atmosphere. Several studies have been carried out to reduce the use of PC in concrete for addressing global warming issues. This has created pressures on the construction industries. An attempt in this regard is the development of geopolymer concrete (GPC), which seems to be the future of construction industries. Geopolymer is an innovative binder and can be produced by treating materials of geological origin or from by-product (rich in silicon and aluminum) with highly alkaline solution [123]. These include clays [124], silica fumes [125], both low-carbon and high-carbon fly ash [126], red mud [127], both high-carbon [128] and low-carbon blast furnace slag [129], rice husk ash [130], zeolites [131]; and argon oxygen decarburization (AOD) slag [132], bottom ash [133], pond ash [134], man-made rock [135], biochar concrete [136], attapulgite pozzolana [137], copper mine tailings [138], activated blast furnace slag [139], low Ca electric arc ferronickel slags, metakaolin, etc., are alternatives to Portland cement [105–108].

Scientists have used different methods, including curing [140], sol-gel [141], deep mixing [142], addition of additives such as Na_2CO_3, $CaCO_3$, $MgCO_3$, hydrothermal synthesis and low-temperature calcination methods, mechanical activation (MA), blended mineral admixtures, an alkaline solution of thermally activated silica-alumina bearing mineral with additives such as calcium carbonate; and innovative *in-situ* co-reticulation process, an innovative method to reclaim the waste molding sands containing water glass with "dry" or "wet" activation of inorganic binder in waste molding sand mixtures physically hardened by microwave radiation, and sintering process [143]. Some related reports follow.

Zheng *et al.* [144] used sol-gel method for preparing geopolymer with pure Al_2O_3-$2SiO_2$ precursors (powders) through alkali-activation process.

Badogiannis *et al.* [145] prepared supplementary cementitious material. They studied optimization of the calcined kaolin, in which samples were heated at different temperatures at different times.

Wei *et al.* [146] fabricated plant fiber-based geopolymer composites using metakaolin, alkaline sodium silicate and plant fibers. Orthogonal test and single factor analysis were used to study the influence of water glass modulus, solid liquid ratio, and fiber content on bending strength. The results showed the influence of factors (solid to liquid ratio, fiber content, water glass modulus) on bending strength.

Ahmari and Zhang [147] prepared eco-friendly bricks utilizing copper mine tailings through geopolymerization technology. They studied the feasibility of copper mine tailings for production of eco-friendly bricks.

Jeon *et al.* [148] explored the preparation of activated geopolymer. They studied the beneficial effects of Na_2CO_3 as an additive for microstructural and strength properties. $Ca(OH)_2$-activated fly ash system and NaOH-activated fly ash samples were also tested by compressive testing, XRD, SEM/BSE/EDS, 29Si/27Al MAS-NMR, MIP and TGA, and compared the effects of Na_2CO_3.

Akhtar and Sarmah [149] used biochar, which is a carbonaceous solid material produced from three different waste sources (poultry litter, rice husk, and pulp and paper mill sludge), during the preparation of concrete. They replaced cement content up to 1% of total volume. The effect of individual biochar mixed with cement on the mechanical properties of concrete was investigated by different characterization techniques.

Khudhair *et al.* [150] aimed to valorize a mineral and natural resources, such as limestone fillers (F-Lime) and natural pozzolan (PN), by introducing them into the formulation of cement and/or concrete matrix. The purpose of taking up F-Lime and PN is to minimize CO_2 emissions, energy consumption, and raw materials.

Vafaei and Allahverdi [151] synthesized geopolymer by alkaline activation using waste-glass powder. They used alkaline solution, which is a mixture of aqueous solutions of sodium hydroxide and sodium silicate, with different Na_2O contents. Three types of calcium aluminate cements were also incorporated into the dry binder at levels up to 24% by weight in order to modify the chemical composition of the geopolymer source materials.

Bouchikhi *et al.* [152] prepared waste glass-based geopolymers using crushed residual waste glass (RWG) and activating solution, which is an eco-friendly inorganic binder. RWG was used as source of free silicon and metakaolin (MK) as source of aluminosilicates.

Roviello *et al.* [153] prepared and characterized geopolymer-based hybrid composites using commercial epoxy-based organic resins (up to 25% in weight of resin) and a metakaolin (up to 75% in weight of resin).

Nguyen Ngoc Lam [154] developed super-sulfated cement (SSC) obtained from phosphogypsum (PG) and ground granulated blast furnace slag (GFS), with a small amount of cement, which is unburnt cementitious material and environmentally friendly.

Gijbels *et al.* [155] prepared hardened binder using portlandite and ettringite intermixed with the nardite and minor amounts of secondary gypsum and merwinite.

Dhavamani Doss *et al.* [156] prepared high-strength alkaline-activated concrete structures using low-calcium fly ash, Ground granulated blast furnace slag (GGBS), metakaolin, and manufacturing sand (mix ratio of 1:1.31:2.22) and used to build massive structures such as sky scrapers, bridges, tunnels, nuclear plants, underground structures. Water to cement ratio (for high-strength cement concrete) and alkaline solution to binder ratio (for alkaline-activated concrete) is 0.35.

Phair *et al.* [157] established aluminosilicate hydrogels as a model system for processing and developing commercial geopolymer binders.

Gong *et al.* [158] manufactured fly ash belite cement (FABC) composed of α0L-C2S and C12A7. The FABC is manufactured using low-grade fly ashes through hydrothermal process followed by low-temperature calcination.

Kalinkin *et al.* [159] synthesized geopolymers at ambient temperature. Blends of fly ash and natural calcite were mechanically activated for 0–400 s in a planetary mill. The calcite content in the blends was 0–10 wt%. NaOH (sodium hydroxide) solution is used as an alkaline agent.

Aboulayt *et al.* [160] prepared geopolymer in an alkaline solution using thermally activated kaolinite clay and calcium carbonate mineral (as an additive).

Rao *et al.* [161] prepared complex binders containing two or three blended mineral admixtures in terms of glass powder (GP), limestone powder (LP), and steel slag powder (SP).

Gupta *et al.* [162] developed hybrid inorganic-organic geopolymer having Si-O-Al networks using fly ash, sodium hydroxide along with rice husk.

Kastiukas *et al.* [163] prepared geopolymer, which is a type of promising alternative binder. They cured the geopolymer in two ways: oven radiation and microwave radiation.

Provis *et al.* [164] synthesized geopolymers through alkali-silicate activation, which is a class of aluminosilicate materials cured in ambient or higher temperature. Powder X-ray diffraction, microscopy, electron diffraction, mechanical strength testing, and calorimetry were used to correlate structure properties.

Sawan *et al.* [165] prepared non-colored and colored geopolymers. Nano nickel aluminate (10 wt%) is taken as colorizing agent for colored geopolymer, whereas dried non-colored geopolymer is sintered at 800–1200 °C.

Zhang *et al.* [166] manufactured geopolymer foam concretes (GFCs) in the laboratory with 0–16% foam additions.

Kargin *et al.* [167] synthesized geopolymer using fly ash. During the geopolymerization process, hydrogen peroxide (H_2O_2) solution was used as the foaming agent. The activation time and the temperature of isothermal holding and hardening under normal conditions for all samples were constant.

Chindaprasirt and Rattanasak [168] prepared fly ash-based geopolymer which requires longer heat curing.

Sayieda *et al.* [169] prepared geopolymer resin by alkali activation of ceramic waste (AACW). The AACW was replaced by 10 and 30% by weight (wt.) concrete of waste (CoW) and 10 and 30 wt% ground granulated blast-furnace slag (GGBFS).

Komnitsas *et al.* [170] prepared geopolymer using low-Ca electric arc ferronickel slags. The additives, such as kaolinite, sodium silicate, sodium hydroxide and water, are used during the geopolymer polymerization process.

Sokołowska [171] prepared polymer composite concrete with filler and investigated durability and long-term compressive strength of polymer concrete containing common and alternative fine fillers, including quartz powder (ground sand) and by-products of the combustion of Polish fossil fuels (coal and lignite).

Chandrasekhar and Pramada [172] used kaolin as cheap raw material, which was used for synthesizing zeolite. It involved a two-step reaction, i.e., dehydroxylation of kaolin (at 550–900 °C) and hydrothermal treatment of metakaolin with aqueous alkali.

Stachowicz and Granat [173] prepared an innovative binder to reclaim the waste molding sands containing water glass, which is hardened by microwave radiation. The sand mixtures consisting of high-silica sand and water-glass with average molar module were subjected to the following cyclical process, viz., mixing the components, compacting, microwave heating, cooling-down, thermally loading the mold to 800 °C, cooling-down to ambient temperature, and knocking-out. The waste moldings sands were subjected to either dry or wet activation of the binder.

Sarmin [174] prepared wood with eco-friendly binder composites, which are widely used in consumer products, i.e., structural and non-structural applications. They developed binder which was environmentally friendly and had enhanced sustainability. Such binder materials are known as geopolymers (i.e., alumino-silicates).

Tantawy *et al.* [175] described low-temperature synthesis of belite (β-C2S) from silica fume with requisite amounts of lime and $BaCl_2$, which were treated hydrothermally in a stainless-steel capsule at 110–150 °C for 2–5 h followed by calcination at 600–700 °C for 3 h to form belite.

Ferone *et al.* [176] developed metakaolin-based geopolymeric mortar to be used as bonding matrix for external strengthening of reinforced concrete beams. During the formation of the geopolymer, they varied the composition of the activating solution in terms of SiO_2/Na_2O ratio.

Sakkas *et al.* [177] developed fire-resistant geopolymers. Their performance was examined under thermal loading. The geopolymers were prepared using metallurgical slag, metakaolin and highly alkaline potassium hydroxide. Geopolymer paste was obtained by curing at 70 °C for a certain period of time.

Mobasher *et al.* [178] prepared the binder structure of sodium sulfate-activated slag cements. The behavior of the materials depends on sodium sulfate content and curing duration. Ettringite, a calcium aluminum silicate hydrate (C-A-S-H) phase, and a hydrotalcite-like Mg-Al layered double hydroxide were identified in the main reaction products. Curing had no effect on the products. The most significant changes on strength properties were observed upon curing at advanced ages.

Suz-Chung Ko [179] described an activated supersulphated aluminosilicate binder containing aluminosilicate, calcium sulphate and an activator containing alkali metal salts. The selected group (i.e., aluminosilicate) is obtained from blast furnace slag, clay, marl and industrial by-products (such as fly ash).

Mohammed and Saeed [180] produced geopolymer concrete, which is a new construction material in Iraq. Foreign fly ash and local fly ash obtained from the south Baghdad Power Plant were used with good specifications to obtain geopolymer concrete. In addition, local pozzolana material was used to explore the possibility of production of geopolymer concrete.

Romero *et al.* [181] produced glass foams formed by alkali activation and gel casting. The alkali activation was done by soda-lime waste glass powders, which allowed obtaining a well-dispersed concentrated suspension. It underwent gelification by curing at low temperature (75 °C).

Purgstaller *et al.* [182] prepared temporal amorphous stabilizer by taking up carbonate mineral. It is prepared either in biotic environments or in abiotic environments.

Rademan *et al.* [183] prepared magnesium oxy chloride (sorel)-based cements and magnesium oxy sulfate-based cements, which have water and corrosion resistance. They incorporated various alkali metal phosphates, such as $MgHPO_4$ or MgH_2PO_4, with alkali metal fatty acids (magnesium stearate; and metal or alkali metal sulfates such as aluminum sulfate or magnesium sulfate). Moreover, water resistance is enhanced by either pre-carbonating the mix water or the liquid magnesium chloride phase of the cements, or by adding a carbonate into the powder phase. Accelerated cure of this system has also been obtained by using various inorganic metal oxides.

Azimi *et al.* [184] used fly ash and dolomite as raw materials during the production of geopolymer composite using alkaline solution (sodium hydroxide, and liquid sodium silicate).

Lee and Soh [185] developed carbon negative cement. This is one of the most typical methods for reducing CO_2 in building materials, either by addition of slag and fly ash, like pozzolana material, or by developing carbon negative cement.

1.7.2 Geopolymer Properties

In the development of geopolymer framework, various studies have been performed in order to determine the physical and chemical properties, and durability of geopolymers. Physical properties, such as temperature, electric or magnetic field, or light and chemical properties of materials, are subjected to a more or less aggressive environment. Other properties, i.e., mechanical properties, reflect the performance of materials, which is deformed by force systems.

Geopolymer products possess high strength, low shrinkage, and high resistance to freezing, thawing, and corrosion. Compressive strength is found to be 20 MPa after 4 h curing at 20 °C and 70–100 MPa after 28 days curing [186]. Compressive strength of geopolymer mortar (cube-shaped) is found to be 40 MPa after 28 days curing [187].

Thermal energy is an accelerator of geopolymer cements. Since, thermal energy is a function of temperature, geopolymer products gain mechanical strength upon curing. Generally, temperature and other types of activators are important parameters that affect the strength properties of geopolymer products [188].

The heat and fire resistance of geopolymer binder-based products is shown to be better than Portland cement-based products [186, 189].

At high temperature (up to 300 °C), OPC-based products promptly deteriorate in compressive strength, but geopolymer cements are stable at 600 °C. Also, geopolymer binders have extremely low shrinkage compared to Portland cement [190].

Other than physical and mechanical properties, geopolymers have excellent chemical properties. Both geopolymer pastes and mortars have been proven to have satisfactory corrosion behavior in different corrosive medium, i.e., sulfates, seawater attack, acidic media, and alkali silica reaction [191].

The standard accelerated mortar bar test is used to demonstrate the alkali-aggregate resistance of geopolymer paste compared to OPC-based products. Results were very satisfactory for geopolymer pastes [192].

The most interesting behavior of geopolymer binder is its resistance to acid corrosion. Alkali activated binder showed better corrosion results than OPC-based products [193]. In HCl corrosive media, alkali-activated fly ash appeared to be healthier after 90 days than OPC-based specimens [194].

1.8 Durability of Concrete

In 1930, the deterioration of concrete was attributed either to crumbling or leaching at joints or poorly consolidated concrete. In the period from 1950 to 1980, the changes in the composition and hydration characteristics of Portland cement-based concrete affected the durability of concrete structure. Then, in the 21st century, civil engineers began facing the challenge of building concrete structures that were more environmentally sustainable.

The deterioration process of concrete structures can result in cracking [195]. Generally, cracking in concrete structures is due to the mechanical or environmental loading. Cracks may happen in non-structural concrete due to sedimentation, shrinkage, or thermal movement. However, concrete structural elements have excessive tensile stress. There is a correlation between cracks and the durability of concrete structures. Since crack development in concrete structures can affect the strength, stiffness, and long-term durability of concrete-based structures [196], crack formation and crack propagation can be controlled by improving the durability of concrete structures.

The mechanism of steel corrosion in reinforced concrete is an important subject. It causes the deterioration of concrete structure in seawater environment through electrochemical process. In the electrochemical process, a cathode receives electrons from an anode (Figure 1.25). Transfer of electrons between different regions of the steel reinforcement will occur. One area will behave as anode and the other as cathode. Thus, steel corrosion in aqueous solution requires two elementary electrochemical reactions: steel oxidation and reduction of oxygen dissolved in solution. The immediate consequences of rusting of rebar are cracking, spalling or delamination of concrete surface. Consequently, all these elements lead to an easier ingress of aggressive agents, and then accelerate corrosion rate. Stages of corrosion-induced deterioration (structural preservation systems) are shown in Figure 1.26.

Pores of concrete structures in a marine environment create a serious problem for long-term stability, i.e., durability. The pore solution of

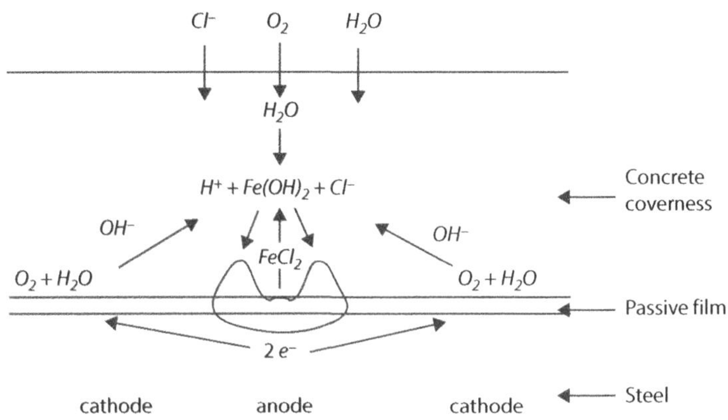

Figure 1.25 Dissolution mechanism of steel in concrete due to chloride ion ingress.

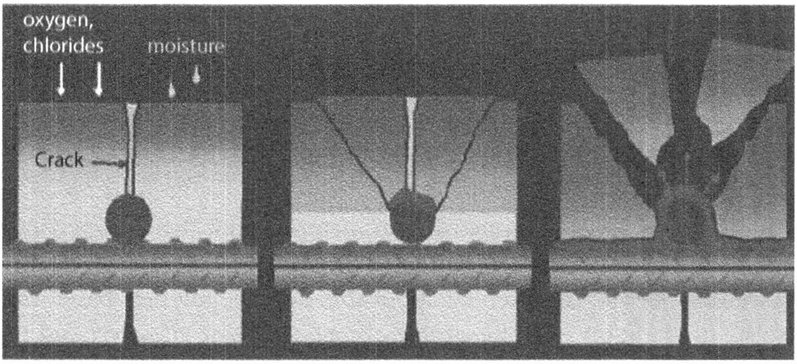

Figure 1.26 Stages of corrosion-induced deterioration (structural preservation systems) [197].

concrete may reach an alkalinity level (pH 13). Thus, this alkaline solution creates an oxide layer in the presence of oxygen on steel surface and protects embedded steel from corrosion in an efficient manner.

This system of protection is known as passivity of embedded steel. However, the integrity and protective quality of this film are dependent on factors such as oxygen availability and alkalinity of the solution. Generally, corrosion of steel reinforcement in aqueous media involves reactions between the metal and the solution. The contribution of this interface to the corrosion phenomena is preponderant [198].

1.9 Accelerated Durability Testing

Ahmad [199] has discussed the corrosion property of steel-reinforced concrete structures. This behavior is a slow process, which may be due to the formation of a protective layer on concrete.

Kasai and Nakamura [200] studied the durability behavior of cement mortars in seawater media through an accelerated test method. Results displayed the corrosion process of the test specimens. Deterioration of cement mortar is greater in high drying temperature and high water-cement ratio.

Nishibayashi *et al.* [201] reported the accelerated testing process using cement-based concrete. In the test method, the dynamic modulus of elasticity and the change in length are estimated. They concluded that the specimens made with seawater resistant cement and containing about 80% of blended blast furnace slag performed better than OPC concrete. Moreover, they estimated that this accelerated testing method provided acceptable results.

1.10 Conclusion(S)

Chronological studies on construction materials and their properties are being extensively studied due to their potential applications in construction sectors. For the commercial applications of geopolymer-based materials, the preparation must be simple, direct, and cost-effective; and must also improve material properties.

Acknowledgments

The authors convey their sincere thanks to the Ministry of Mines, Government of India, for providing a grant to carry out the work [Grant number F.No.:14/54/20214-Met.-IV dated: 29.12.2014]. They also convey their sincere thanks to GIET, Gunupur, Rayagada, Odisha, India, for providing lab facilities to do the research work. The authors also would like to thank the CRF, IIT Kharagpur, for providing their testing facilities.

References

1. History of construction, From Wikipedia, the free encyclopedia.
2. Alfred Swenson and Pao-Chi Chang, "History of Building".
3. Atkinson, Richard, Stonehenge Penguin Books, 1956.
4. A paper showing the joints used at Stonehenge.
5. A. Johnson, Solving Stonehenge: The New Key to an Ancient Enigma. (Thames & Hudson, 2008), ISBN 978-0-500-05155-9.
6. T.S. Kawami, Parthian Brick Vaults in Mesopotamia, Their Antecedents and Descendants, Columbia University,. *Archived from* the original *on 2013-10-29, Retrieved 2014-04-02.*
7. Alfred Swenson and Pao-Chi Chang (Britannica), The History of Building Construction.
8. N. Hunt, Living in ancient Greece. New York, N.Y.: Chelsea House Publishers, 2009. 24. ISBN 0816063397
9. C. Strickland and A. Handy. The Annotated Arch: A Crash Course in History of Architecture. Kansas City, MO: Andrews McMeel Pub., 2001. 12. ISBN 0740710249
10. "The History of Building".

11. Stephany, Erich Der Dom zu Aachen (Aachen Cathedral) Arend und Ortmann, Aachen, 1972.
12. Upton, Dell. Architecture in the United States. Oxford: Oxford University Press, 1998. 153. ISBN 019284217X.
13. Bill Addis. Building: 3000 years of Design Engineering and Construction. Phaidon. 2007. p. 632
14. A. Becchi, M. Corradi, F. Foce and O. Pedemonte (Eds.), Construction History: Research Perspectives in Europe. Associazione Eduardo Benvenuto. 2004
15. http://www.constructionhistory.co.uk
16. First International Congress on Construction History Archived 2011-07-16 at the Way back Machine Madrid 2003.
17. Second International Congress on Construction History Archived 2011-09-30 at the Way back Machine'at University of Cambridge, 2006.
18. Third International Congress on Construction History Archived 2012-07-24 at the Wayback Machine May 2009
19. Fourth International Congress on Construction History Paris, July 2012.
20. E.K. Serafimova, M. Mladenov, I. Mihailova, Y. Pelovski, (2011), Study on the Characteristics of Waste Wood Ash, Journal of the University of Chemical Technology and Metallurgy, Vol. 46, pp. 31-34.
21. M.K. Mishra, K.W. Ragland, and A.J. Baker†, (1993), Wood Ash Composition as a Function of Furnace Temperature, Biomass and Bioenergy Vol. 4, pp. 103-116.
22. G.A. Habeeb and H.B. Mahmud, (2010), Study on Properties of Rice Husk Ash and Its Use as Cement Replacement Material, Materials Research, Vol. 13, pp. 185-190.
23. SK S. Hossain, L. Mathur, and P.K. Roy, (2018), Rice husk/rice husk ash as an alternative source of silica in ceramics: A review, Journal of Asian Ceramic Societies, Vol. 6, pp.299-313.
24. F. Prata and B. Rose, in The Encyclopedia of Volcanoes (Second Edition), 2015, pp. 911-934.
25. Claire J. Horwell . Peter J. Baxter, The Respiratory Health Hazards of Volcanic Ash: A Review for Volcanic Risk Mitigation-Review Article, Bulletin of Volcanology pp.1-25 (DOI 10.1007/s00445-006-0052-y).
26. W.I. Rose and A.J. Durant, (2009), Fine Ash Content of Explosive Eruptions, Journal of Volcanology and Geothermal Research, Vol. 186, pp.32-39.
27. K.V. Cashman, B. Sturtevant, P. Papale, O. Navon, (2000), Magmatic Fragmentation, In Sigurdsson, H.; Houghton, B.F.; McNutt, S.R.; Rymer, H.; Stix, J. (Eds.), Encyclopedia of Volcanoes. San Diego, USA: Elsevier Inc. p. 1417.
28. K.S. Prakash and Ch. Hanumantha Rao, (2016), Study on Compressive Strength of Quarry Dust as Fine Aggregate in Concrete, Advances in Civil Engineering, Vol. 2016, pp. 1-6 (http://dx.doi.org/10.1155/2016/1742769).

29. V.K. Nagarajan, S.A. Devi, S.P. Manohari, M.M. Santha, (2014), Experimental Study on Partial Replacement of Cement with Coconut Shell Ash in Concrete, International Journal of Science and Research (IJSR), Vol. 3, pp.xxx-xxx.
30. Utsev, J. T., Taku, J. K., (2012), Coconut Shell Ash As Partial Replacement of Ordinary Portland Cement In Concrete Production, International Journal of Scientific and Technology Research, Vol. 1, pp. xxx-xxx.
31. O.M. Ikumapayi, E. Akinlabi, J.D. Majumdar, S. Akinlabi, (xxx), Applications of Coconut Shell Ash/Particles in Modern Manufacturing: A Case Study of Friction Stir Processing, In book: Modern Manufacturing Processes, Chapter: 4, Publisher: Elsevier Wood head Publishing publications (DOI: 10.1016/B978-0-12-819496-6.00004-X).
32. Ash, From Wikipedia, the free encyclopedia
33. KGR Agro Fusions (P) Limited, Ludhiana, Punjab
34. Rice Husk Ash in Concrete: Uses, Pros & Cons & Its Effect on Properties of Concrete, Conveyor Concrete Placing National Leader in Conveyor Concrete Placement, Bridges, Dams, Foundations, etc., Rotec Industries.
35. Cigars 101, Explore the World of Cigars, Holts Club House.
36. EDT, 'Volcanic Ash' Isn't Actually Ash, Ethan Siegel, Senior Contributor, Starts with A Bang, Contributor Group, Science.
37. Sri Srinivas Stone Crusher, Vadapalani, Chennai, Tamil Nadu and Krishna Suppliers, West Mambalam, Chennai, Tamil N. Indiamart.com.
38. Omolayo M. Ikumapayi, 1Esther T. Akinlabi, 1Jyotsna D. Majumdar, 2Stephen A. Akinlabi3, Chapter four - Applications of coconut shell ash/particles in modern manufacturing: a case study of friction stir processing Modern Manufacturing Processes Woodhead Publishing Reviews: Mechanical Engineering Series, 2020, Pages 69-95
39. World oil outlook, OPSC, Secretariate background paper, 15th International Energy Fourm ministeriate, Algieris, Algeria, 26th-28th Sep 2016.
40. Suravi, (2017), Fly ash in India: Generation vis-à-vis Utilization and Global prospective, International Journal of Applied Chemistry, Vol. 13, pp. 29-52.
41. Coal Fly Ash -User Guidelines for Waste and Byproduct Materials in Pavement Construction- Material Description- FHWA-RD-97-148.
42. American coal ash Association-Fly Ash Facts for Highway Engineers-Chapter 1 - Fly Ash - An Engineering Material-FHWA-IF-03-019.
43. R Sokolar and M Nguyen, (2018), The Fly Ash of Class C for Ceramic Technology, IOP Conf. Series: Materials Science and Engineering, Vol. 385, pp. 012053-xxx. (DOI:10.1088/1757-899X/385/1/012053 Construmat 2018).
44. A Wardhono, (2017), Comparison Study of Class F and Class C Fly Ashes as Cement Replacement Material on Strength Development of Non-Cement Mortar, IOP Conference Series: Materials Science and Engineering, Vol. 288 (The 2nd Annual Applied Science and Engineering Conference (AASEC 2017) 24 August 2017, Bandung, Indonesia).

45. M.K. Panigrahi, (2021) Investigation of Structural, Morphological, Resistivity of Novel Electrical Insulator: Industrial Wastes, Bull. Sci. Res., Vol. 3, pp. 51-58.
46. M. Panigrahi, R.R. Dash, R.I. Ganguly, Optimization Mechanical Properties of Pond Ash Geopolymer: As Construction Material" in National Conference on Advanced Engineering Materials-2016 (ISBN No. 978-93-81693-07-3, 23-24th July 2016).
47. US Federal Highway Administration. (July 17, 2015), User Guidelines for Waste and By-product Materials in Pavement Construction (Available: http://www.fhwa.dot.gov/ publications/ research/infrastructure/structures/97148/cbabs1.CFM)
48. L.M. Deraman, M.M. Al Bakri Abdullah, L.Y. Ming, and K. Hussin, (2017), Density and Morphology Studies on Bottom Ash and Fly Ash Geopolymer Brick, AIP Conf. Proc. 1835, 020047-1–020047-6 (DOI: 10.1063/1.4981869 Published by AIP Publishing. ISBN: 978-0-7354-1505-8, pp. 020047-6.
49. U.S. Environmental Protection Agency (EPA), Washington, D.C. "Hazardous and Solid Waste Management System; Identification and Listing of Special Wastes; Disposal of Coal Combustion Residuals from Electric Utilities." Proposed rule. Federal Register, 75 FR 35130, June 21, 2010 (https://www.federal register.gov/citation/75-FR-35130).
50. C. Erickson, (2019), Mixing water, Powder River Basin Coal Ash Dangerous Tohuman Health, New Research Finds, Casper Star-Tribune. Casper, WY. (https://trib.com/business/energy/mixing-water-powder-river-basin-coal-ash-dangerous-to-human/article_9a85a08e-8d70-5dad-b059-b861d2a34757.html).
51. Effluent Limitations Guidelines and Standards for the Steam Electric Power Generating PointSource Category" EPA. 2018-11-30 (https://www.epa.gov/eg/steam-electric-power-generating-effluent-guidelines-201 5-final-rule).
52. Brooke, Nelson (June 5, 2019). "New Interactive Maps of Groundwater Pollution Reveal ThreatsPosed by Alabama Power Coal Ash Pits". Black Warrior Riverkeeper. Birmingham, AL. (https://blackwarriorriver.org/new-coal-ash-pollution-map s/)
53. Springer, Patrick (March 13, 2019). "Report: Unsafe coal ash contamination found in North Dakotagroundwater". Bismarck Tribune. Bismarck, ND. (https://bismarcktribune.com/news/state-and-regional/report-unsafe-coal-ash-contami nation-found-in-north-dakota-groundwater/article_ccb-da15d-5489-55f0-a756-84ebb7dd062f.html)
54. T. Tosheff, (2019-07-31). "York: Brunner Island Power Plant Owners Agree to $1M Penalty, Coal Ash Cleanup" Harrisburg, PA: ABC27 News (https://www.abc27.com/news/local/york/brunner-island-power-plant-owners-agree-to 1m-penalty-coal-ash-cleanup).
55. S. Dewan, (2008-12-23), Water Supplies Tested after Tennessee Spill, The New York Times (https://www.nytimes.co m/2008/12/24/us/24mud.html).

56. "Case Summary: Duke Energy Agrees to $3 Million Cleanup for Coal Ash Release in the Dan River" Enforcement. EPA. 2017-03-15 (https://www.epa.gov/enforcement/case-summary-duke-energy-agrees-3-million-cleanup-coal-ash-r elease-dan-river)
57. P. Ghosh and S. Goel, (2014), Physical and Chemical Characterization of Pond Ash, International Journal of Environmental Research and Development. Vol. 4, pp. 129-134.
58. C.A. Langton, N. Rajendran, S.E. Smith, (1998), Use of Pond Ash in CLSM, Concr Int., Vol. 20, pp. 58-62.
59. T.R. Naik and R.N. Kraus, (1999), The Role of Flowable Slurry in Sustainable Developments in Civil Engineering, Proc., ASCE Conf. on Materials and Construction—Exploring the Connection, ASCE, Reston, Va., pp. 826-834.
60. Langton CA, Rajendran N, Smith SE (1998) Use of pond ash in CLSM. Concr Int 20:58–62.
61. A. Bhatt, S. Priyadarshini, A.A. Mohanakrishnan, A. Abri, M. Sattler, S. Techapaphawit, (2019), Physical, Chemical, and Geotechnical Properties of Coal Fly Ash: A Global Review Author Links Open Overlay Panel, Case Studies in Construction Materials, Vol. 11, pp. 00263-xxx (https://doi.org/10.1016/j.cscm.2019.e00263).
62. S.M. Rao and I.P. Acharya, (2014), Synthesis and Characterization of Fly Ash Geopolymer Sand, Journal Of Materials In Civil Engineering ASCE, Vol. 26, pp.912-917.
63. R. Kumar, (2010), Mineralogical, Chemical and Morphological Studies of Fly Ashes from Thermal Power Stations of India, Asian Journal of Water, Environment and Pollution, Vol. 7, pp. 103-110.
64. J. Feng, J. Sun, and P. Yan, (2018), The Influence of Ground Fly Ash on Cement Hydration and Mechanical Property of Mortar, Advances in Civil Engineering, Vol. 2018, pp. 7.
65. K. ZabielskaAdamska, (2020) Hydraulic Conductivity of Fly Ash as a Barrier Material: Some Problems in Determination, Environmental Earth Sciences, Vol. 79, pp.321-333.
66. H. Li, G. Liu, and Y. Cao, (2014), Content and Distribution of Trace Elements and Polycyclic Aromatic Hydrocarbons in Fly Ash from a Coal-Fired CHP Plant, Aerosol and Air Quality Research, Vol. 14, pp. 1179-1188.
67. G.A. Leonards and B. Bailey, (1982), Pulverized Coal Ash as Structural Fill', J. Geotech. Eng., ASCE, Vol. 108, pp.517-531.
68. Coal Physical Testing, SGS Minerals Services-T3 SGS 527, 10-2013. Email us at minerals@sgs.com www.sgs.com/coal
69. A. Sridharan and K. Prakash, (2000), Classification Procedures for Expansive Soils, Proceedings of the Institution of Civil Engineers Geotechnical Engineering, Vol. 143, pp. 235-240.
70. A. Trivedi and V. K. Sud, (2002), Grain Characteristics and Engineering Properties of Coal Ash, Granular Matter, Vol. 4, pp. 93-101 (DOI 10.1007/s10035-002-0114-6).

71. W.R. Roy and P.M. Berger, (2011), Geochemical Controls of Coal Fly Ash Leachate pH, Coal Combustion and Gasification Products, Vol. 3 pp.63-66.
72. P. Trinh, B. Yuko, O. Kenichiro, and N.K. Kawai, (2015), A Study on Pozzolanic Reaction of Fly Ash Cement Paste Activated by an Injection of Alkali Solution, Construction and Building Materials, Vol. 94, pp. 28-34.
73. N. Pandian, A. Sridharan, and S. Srinivas, (2001), Compaction behaviour of Indian coal ashes, Proceedings of the Institution of Civil Engineers Ground Improvement, Vol. 5, pp. 13-22.
74. T. Xie and T. Ozbakkaloglu, (2015), No Access Influence of coal ash properties on compressive behaviour of FA- and BA-based GPC, Magazine of Concrete Research, Vol. 67, pp. 1301-1314.
75. S.R. Kaniraj and V. Gayathri, (2004), Permeability and Consolidation Characteristics of Compacted Fly Ash, Journal Energy Engineering, Vol. 130, pp. 18-43.
76. M.R. Hajarnavis, A.D. Bhide, (2003) Leaching Behaviour of Coal-Ash: A Case Study, Indian J Environ Health, Vol. 45, pp. 293-8.
77. Chapter 6, Industrial Solid Waste, National Waste Management Council-Ministry of Environment & Forests-1990/1999)
78. K. Sobolev, High Performance Cement for High Strength and Extreme Durability, Archived from the original on 2009-08-03 (Retrieved 2009-06-18).
79. Generation of iron and steel slag: Nippon Slag Association (**Error! Hyperlink reference not valid.**)
80. Nippon Slag Association conducts investigations, research, and promotion
81. Steel Furnace Slag, Australasian (iron & steel) Slag Association (asa-inc.org.au)
82. Electric Arc Furnace Slag (EAFS), Australasian (iron & steel) Slag Association, asa-inc.org.au.
83. J. Xiang, Q. Huang, G. Pei, X. Lv, S. Liu, and W. Lv, (2018), Mineralogical Characterization and Magnetic Separation of Vanadium-Bearing Converter Slag, Waste Management and Research, Vol. 36, pp. 0734242X1879620 (DOI: 10.1177/ 0734242X18796201).
84. Nippon Slag Association conducts investigations, research, and promotion related to iron and steel slag products.
85. User Guidelines for Waste and Byproduct Materials in Pavement Construction, Federal Highway Administration, Publication Number: FHWA-RD-97-148.
86. Pro Quest Dissertations and Theses; Thesis (Ph.D.)-Purdue University, (2017; Publication Number: AAT 10623787; ISBN: 9780355611571; Source: Dissertation Abstracts International, Vol. 79-07(E), Section: B.; 132 p.2017 (AAT 10623787)
87. Ground Granulated Blast-Furnace Slag, Wikipedia, the free encyclopedia.
88. S. Chand, B. Paul, and M. Kumar, (2016), Sustainable Approaches for Ld. Slag Waste Management in Steel Industries: A Review, Metallurgist Vol. 60, pp 116- 128. (DOI 10.1007/s11015-016-0261-3).

89. Indian Minerals Yearbook 2015 (Part- II: Metals & Alloys) 54th Edition Slag-Iron and Steel, Government of India Ministry of Mines Indian Bureau of Mines Indira Bhavan, Civil Lines, NAGPUR-440 001.
90. Nippon Slag Association conducts investigations, research, and promotion related to iron and steel slag products, About iron and steel slag.
91. S. Eloneva, S. Teir, J. Salminen, C.-J. Fogelholm, and R. Zevenhoven, (2008), Steel Converter Slag as a Raw Material for Precipitation of Pure Calcium Carbonate, Ind. Eng. Chem. Res., Vol. 47, pp. 7104-7111.
92. H. Moon, K. Kim, J.-H. Kim, M. Lee, and C.-W. Chung, (2021), Reaction of Converter Slag with Supercritical Carbon Dioxide and Its Potential Applicability as Aggregate for Concrete, Appl. Sci., Vol. 11, pp.1918-15. (https://doi.org/10.3390/app11041918).
93. Electric Arc Furnace Slag Retrieved from https://encyclopedia.pub/3243.
94. Chemical characteristics of iron and steel slag Nippon Slag Association conducts investigations (Research, and promotionrelated to iron and steel slag products).
95. I.Z. Yildirim and M. Prezzi, (2011), Chemical, Mineralogical, and Morphological Properties of Steel Slag Advances in Civil Engineering, Vol. 2011, pp.1-14 (DOI:10.1155/2011/463638).
96. J. Davidovits, (1978), Process for Agglomerating Compressible Mineral Substances in the Form of Powder, Particles and Fibres: United Kingdom Patent UK 1,518,605.
97. J. Davidovits, (2017), Review Geopolymers: Ceramic-Like Inorganic Polymers, J. Ceram. Sci. Technol. 08 [3] 335-350 (2017 (DOI: 10.4416/JCST2017-00038).
98. E.P.Ivanova Kateryna, B. Russell and J. Crawford, (2014), Chapter 4 - Advanced Synthetic and Hybrid Polymer Biomaterials Derived from Inorganic and Mixed Organic-Inorganic Sources, New Functional Biomaterials for Medicine and Healthcare, Vol. 2014, pp. 100-120.
99. P. Jittabut and S. Horpibulsuk, (2019), Physical and Microstructure Properties of Geopolymer Nanocomposite Reinforced with Carbon Nanotubes, Materials Today proceedings, Vol. 17, pp. 1682-1692.
100. V.D. Glukhovsky, (1959), Soil silicates, Gosstroyizdat, KIEV, pp. 154
101. D. Khale and R. Chaudhary, (2007), Review on Mechanism of Geopolymerization and Factors Influencing Its Development: A Review, J Mater Sci., Vol. 42, pp.729-746 (DOI 10.1007/s10853-006-0401-4).
102. M.M.A. Abdullah, K. Hussin, M. Bnhussain, K.N. Ismail, and W.M.W. Ibrahim, (2011), Mechanism and Chemical Reaction of Fly Ash Geopolymer Cement- A Review, International Journal of Pure and Applied Sciences and Technology, Vol. 6, pp. 35-44.
103. H. Rahier, J. Wastiels, M. Biesemans, R. Willlem, G. Van Assche, and B. Van Mele, (2007), Reaction Mechanism, Kinetics and High Temperature Transformations of Geopolymers, J Mater Sci., Vol. 42, pp. 2982-2996 (DOI 10.1007/s10853-006-0568-8).

104. P. Cong and Y. Cheng, (2021), Advances in Geopolymer Materials: A Comprehensive Review, Journal of Traffic and Transportation Engineering (English Edition), Vol. 8, pp. 283-314.
105. Phair, J.W. (2006), Green Chemistry for Sustainable Cement Production and Use, Green Chem., Vol. 8, pp. 763-780.
106. U.S. Geological Survey. Mineral Commodity Summaries 2018; U.S. Geological Survey: Reston, VA, USA, 2018.
107. R.A. Smith, J.R. Kersey, and P.J. Griffiths, (2002), The Construction Industry Mass Balance: Resource Use, Wastes and Emissions, Construction, Vol. 4, pp. 680-xxx.
108. A. Hasanbeigi, L. Price, H. Lu, and W. Lan, (2010), Analysis of Energy-Efficiency Opportunities for the Cement Industry in Shandong Province, China: A case study of 16 cement plants. Energy, Vol. 35, pp. 3461-3473.
109. L.K. Turner and F.G. Collins, (2013), Carbon Dioxide Equivalent ($CO_{2-}e$) Emissions: A Comparison Between Geopolymer and OPC Cement Concrete, Constr. Build. Mater., 43, 125–130.
110. M. Valipour, M. Yekkalar, M. Shekarchi, and S. Panahi, (2014), Environmental Assessment Of Green Concrete Containing Natural Zeolite On The Global Warming Index In Marine Environments, J. Clean. Prod., Vol.65, pp. 418-423.
111. P. Lund, (2007), Impacts of EU Carbon Emission Trade Directive on Energy-Intensive Industries, Indicative micro-economic analyses. Ecol. Econ. Vol. 63, pp. 799-806.
112. World Business Council for Sustainable Development. The Cement Sustainability Initiative, Cement Industry Energy and CO_2 Performance: "Getting the Numbers Right"; WBCSD: Geneva, Switzerland, 2009; Vol. 44, ISBN 978-3-940388-48-3.
113. A. Bisarya, R.K.Chouhan, M. Mudgal, S.S.Amritphale, (2015), Fly Ash based Geopolymer Concrete a New Technology towards the Greener Environment: A Review, International Journal of Innovative Research in Science, Engineering and Technology, Vol. 4, pp.xxx-xxx.
114. K. Pimraksaa, P. Chindaprasirt, A. Rungchet, K. Sagoe-Crentsil, and T. Sato, (2011) Lightweight geopolymer made of highly porous siliceous materials with various Na_2O/Al_2O_3 and SiO_2/Al_2O_3 ratios, Materials Science and Engineering A, Vol. 528, pp. 6616-6623.
115. P. Nuaklong, A. Wongsa, V. Sata, K. Boonserm, J. Sanjayan, and P. Chindaprasirt, (2019), Properties of High-Calcium and Low-Calcium Fly Ash Combination Geopolymer Mortar Containing Recycled Aggregate, Vol. 5, pp. e02513-xxx.
116. R.E. Lyon, A.J. Foden, P.N. Balaguru, J. Davidovits, and M. Davidovics, (1997), Properties of Geopolymer Matrix-Carbon Fiber Composites, Fire and Materials, Vol. 21, pp. 67-73.
117. R.O. Abdel Rahman, R.Z. Rahimov, N.R. Rahimova, and M.I. Ojovan. (2015), Cementitious Materials for Nuclear Waste Immobilization, ISBN 978-1-118-51200-5, Wiley, Chichester pp. 1-232.

118. *L. Almkvist, S. Bai, W. Bastiaens, C. Cau-dit-Coumes, F. Glasser, J. Govaert, (2013).* The Behaviour of Cementitious Materials in Long-Term Storage and Disposal of Radioactive Waste, IAEA-TECDOC-1701, IAEA, 61 p., Vienna (2013)", *iaea.org* (Retrieved 21 February *2021).*
119. M. Davidovits, M. Bruno, and J. Davidovits, (1999), Past and Present Experience on the Use of Carbon-Geopolymer Composite in Formula One and CART Racing Cars, Geopolymer '99 Proceedings, 141-142.
120. J. Davidovits, (2002), 30 Years of Successes and Failures in Geopolymer Applications, Market Trends and Potential Breakthroughs, Geopolymer 2002 Conference, Oct. 28-29, Melbourne, Australia (https://www.geopolymer.org/category/library/technical-papers).
121. PCT patent application publication WO 2004/106705 filed by Porsche AG, 2004.the Geopolymer Institute page (https://www.geopolymer.org/applications/geopolymer-cement).
122. J. Davidovits and J.L. Sawyer, (1985), Early High-Strength Mineral Polymer, US Patent4,509,985, 1985, filed February 22, 1984. The first commercial geopolymer cement was coined Pyrament 2000™ designed for repair and patching operations.
123. D. Hardjito, S.E. Wallah, D.M.J. Sumajouw, and B.V. Rangan, Brief Review of Development of Geopolymer Concrete, George Hoff Symposium, American Concrete Institute, Los Vegas, USA, 25 May 2004.
124. N. Kumari and C. Mohan, (2021), Basics of Clay Minerals and Their Characteristic Properties (DOI: 10.5772/intechopen.97672).
125. V. Srivastava, R. Kumar, V.C. Agarwal, and P. K. Mehta, (xxx), Effect of Silica Fume on Workability and Compressive Strength of OPC Concrete, J. Environ. Nanotechnol. Vol. 3, pp. 32-35 (DOI: 10.13074/jent.2014.09.143086).
126. R.M. Kalombe, V.T. Ojumu, C.P. Eze, S.M. Nyale, J. Kevern, and L.F. Petrik, (2020), Fly Ash-Based Geopolymer Building Materials for Green and Sustainable Development, Materials, Vol. 13, pp. 5699-17; (DOI:10.3390/ma13245699).
127. Y. Li, X. Min, Y. Ke, D. Liu, and C. Tang, (2019), Preparation of Red Mud-based Geopolymer Materials from Mswi Fly Ash and Red Mud by Mechanical Activation, Waste Management, Vol. 83, pp. 202-208.
128. M. Panigrahi, R.I. Ganguly, R.R. Dash, (2018), An Advanced Cured High Carbon Ferrochrome Slag (HCFCS) Geopolymer (GP): A Constructional Materials, IOP Conference Series: Materials Science and Engineering, Vol. 410, pp.1-15.
129. L. Wang, L. Chen, and D.C.W. Tsang, (2020), Contaminated Soil and Ground water Materials, Processes, and Assessment, Chapter 5 - Green remediation by using low-carbon cement-based stabilization/solidification approaches, Vol.xxx, pp. 93-118.
130. S.K. Das, J. Mishra, and Syed M. Mustakim, (2018) Rice Husk Ash as a Potential Source Material for Geopolymer Concrete: A Review, International Journal of Applied Engineering Research, Vol. 13, pp. 81-84.

131. A. Nikolov, I. Rostovsky, and H. Nugteren, (2017), Case study Geopolymer materials based on natural zeolite, Case Studies in Construction Materials, Vol. 6, pp. 198-205
132. Y.-J. Wang, Y.-N. Zeng, J.-G. Li, and Y.-Z. Zhang, (2020), Cementitious Behavior of Argon Oxygen Decarburization Stainless Steel Slag and Its Stabilization on Chromium September, Crystals 10(10), pp. 876-17 (DOI:10.3390/cryst10100876).
133. E. Haq, S. Kunjalukkal, P.A. Licciulli, (2014), Synthesis and Characteristics of Fly Ash and Bottom Ash Based Geopolymers–A Comparative Study, Ceramics International, Vol. 40, pp. 2965-2971.
134. M. Panigrahi, R.R. Dash, R.I. Ganguly, (2018), Development of Novel Constructional Material from Industrial Solid Waste as Geopolymer for Future Engineers IOP Conference Series: Materials Science and Engineering, Vol. 410, pp.1-12.
135. J. Davidovits, (1994), Geopolymers: Man-Made Rock Geosynthetic and the Resulting Development of Very Early High Strength Cement (Page -2), NASTS award, J. Materials Education, Vol.16 (2&3), pp 91-139.
136. A. Akhtar, A.K. Sarmah, (2017), Novel biochar-concrete composites: Manufacturing, characterization and evaluation of the mechanical properties, Science of The Total Environment, Vol. 616-617, pp. 408-416 (DOI: 10.1016/j.scitotenv.2017.10.319).
137. M.B. Diop, L. Molez, A. Bouguerra, A.N. Diouf, and M.W. Grutzeck, (2014), Manufacturing Brick from Attapulgite Clay at Low Temperature by Geopolymerization, Arabian Journal for Science and Engineering volume 39, pp. 4351–4361.
138. X. Tian, H. Zhang, T. Zhang, and C.A. Fernández, (2020), Alkali-activated copper tailings-based pastes: compressive strength and microstructural characterization, Journal of Materials Research and Technology, Vol. 9, pp. 6557-6567.
139. J. Qiu, Y. Zhao, J. Xing, and X. Sun, (2019), Fly Ash/Blast Furnace Slag-Based Geopolymer as a Potential Binder for Mine Backfilling: Effect of Binder Type and Activator Concentration, Advances in Materials Science and Engineering, Vol. 2019, pp.1-12.
140. G. Görhan and G. Kürklü, (2014), The Influence of the NaOH Solution on the Properties of The Fly Ash-Based Geopolymer Mortar Cured at Different Temperatures, Composites Part B: Engineering, Vol. 58, pp. 371-377.
141. G. Zheng, C. Xue-min, W. Zhang, and Z. Tong, (2009), Preparation of geopolymer precursors by sol–gel method and their characterization, Journal of Materials Science, Vol. 44, pp.3991-3996 (DOI: 10.1007/s10853-009-3549-x).
142. H. Canakci, H. Güllü and A. Alhashemy, (2019), Performances of Using Geopolymers Made with Various Stabilizers for Deep Mixing, Materials, Vol. 12, pp. 2542-32, (DOI:10.3390/ma12162542).
143. N.H. Jamil, M.M. Al Bakri Abdullah, F.C. Pa, H. Mohamad, W.M.A.W. Ibrahim, P. Amonpattaratkit, J. Gondro, W. Sochacki, and N. Ibrahim,

(2021), Self-Fluxing Mechanism in Geopolymerization for Low-Sintering Temperature of Ceramic, Materials, Vol. 14, pp. 1325-12 (https://doi.org/10.3390/ma14061325).
144. G. Zheng, C. Xue-min, W. Zhang, and F. Li, (2012), Preparation of Nano-Sized Al_2O_3–$2SiO_2$ Powder By Sol–Gel Plus Azeotropic Distillation Method, Particuology, Vol. 10, pp. 42-45 (DOI:10.1016/j.partic.2011.07.007).
145. E. Badogiannis, G. Kakali and S. Tsivilis, (xxx), Metakaolin as Supplementary Cementitious Material Optimization of Kaolin to Metakaolin Conversion, Journal of Thermal Analysis and Calorimetry Vol. 81, pp. 457-462.
146. S.J. Wei, J.L. Tan, W.L. Lu, L. Ping Liu, S.J. Yu, and G.J. Zheng, (2018), Preparation and Performances of Geopolymer-Based Plant Fiber Composite solid State Phenomena, Vol. 281, pp. 266-271 (https://doi.org/ 10.4028/www.scientific.net/SSP.281.266).
147. S. Ahmari, L. Zhang, (2012), Production of eco-friendly bricks from copper mine tailings through Geopolymerization, Construction and Building Materials, Vol. 29, pp. 323-331,
148. D. Jeon, Y. Jun, Y. Jeong, J.E. Oh, (2015), Microstructural and Strength Improvements through the Use of Na_2CO_3 in a Cementless $Ca(OH)_2$-activated Class F fly ash system, Cement and Concrete Research, Vol. 67, pp. 215-225.
149. A. Akhtar and A.K. Sarmah, (2017), Novel Biochar-Concrete Composites: Manufacturing, Characterization and Evaluation of the Mechanical Properties, Science of The Total Environment, Vol. 616-617, pp. 408-416 (DOI: 10.1016/j.scitotenv.2017.10.319).
150. M. Khudhair, R Hsissou, M.S. Elyoubi, A Essamri, and A. Elharfi, (2018), Effect Of Interaction Between Inorganic And Organic Admixture On Physical And Mechanical Properties Of Cementitious Material Formulated By Combination Of Natural Pozzolan And Limestone Fillers, April 2018, Conference: Les 4éme Journées de l'Environnement et l'Impact de la Pollution (Eau, Air et Sol) sur la Population et le Réchauffement Climatique (JEIPPRC-04).
151. M. Vafaei and A. Allahverdi (2017), High Strength Geopolymer Binder Based on Waste-Glass Powder, Advanced Powder Technology, Vol. 28, pp. 215-222.
152. A. Bouchikhi, M. Benzerzour, N.-E. Abriak, W. Maherzi, and Y. M.-Pajany, Waste Glass Reuse in Geopolymer Binder Prepared with Metakaolin, 3rd International Conference on Bio-Based Building Materials, 26th-28th June 2019, Belfast, UK., AJCE, Special Issue, Vol. 37, pp. 539-544.
153. G. Roviello, C. Menna, O. Tarallo, L. Ricciotti, F. Messina, C. Ferone, D. Asprone, and R. Cioffi, (2017), Lightweight Geopolymer-Based Hybrid Materials, Composites Part B: Engineering, Vol. 128, pp. 225-237.
154. N.N. Lam, (2018), A study on super-sulfated cement using Dinh Vu phosphogypsum, IOP Conference Series Earth and Environmental Science, Vol. 143, pp. 012016-xxx (DOI: 10.1088/1755-1315/143/1/012016).

155. K. Gijbels, R.I. Iacobescu, Y. Pontikes, S. Schreurs, W. Schroeyers, (2019), Alkali-activated binders based on ground granulated blast furnace slag and phosphogypsum, Construction and Building Materials, Vol. 215, pp. 371-380.
156. S. Dhavamani Doss, S. Thirugnanasambandam, P. Murthi, and K. Poongodi, (2020), Development of Alkaline Activated High Strength Concrete using Fly Ash - Ground Granulated Blast Furnace Slag - Metakaolin as Binders and Manufacturing Sand as Fine Aggregate, International Journal of Innovative Technology and Exploring Engineering (IJITEE), Vol. 9, pp. xxx-xxx.
157. J. Phair, J.D. Smith, and J.S.J Van Deventer, (xxx), Characteristics of Aluminosilicate Hydrogels Related to Commercial Geopolymers, Materials Letters, Vol. 57, pp. 4356-4367 (DOI: 10.1016/S0167-577X(03)00325-2).
158. Y. Gong, C. Liu, Y. Chen, (2020), Properties and Mechanism of Hydration of Fly Ash Belite Cement Prepared from Low-Quality Fly Ash, Applied Sciences, Vol. 10, pp. 7026-xxx (DOI:10.3390/app10207026).
159. A.M. Kalinkin, B.I. Gurevich, E.V. Kalinkina, and I.A. Zvereva (2021), Geopolymers Based on Mechanically Activated Fly Ash Blended with Dolomite, Minerals, Vol. 11, pp. 700-xxx (DOI: 10.3390/min11070700).
160. A. Aboulayt, M. Riahi, M.O. Touham, H. Hannache, M. Gomina, and R. Moussa, (2017), Properties of Metakaolin based Geopolymer Incorporating Calcium Carbonate, Advanced Powder Technology, Vol. 28, pp. xxx-xxx (DOI: 10.1016/j.apt.2017.06.022).
161. M. Rao, J. Wei, Z. Gao, W. Zhou, Q. Li, and S. Liu, (2016), Study on Strength and Microstructure of Cement-Based Materials Containing Combination Mineral Admixtures, Advances in Materials Science and Engineering, Vol. 2016, pp. 1-11 (http://dx.doi.org/10.1155/2016/7243670).
162. R. Gupta, P. Bhardwaj, K. Deshmukh, D. Mishra, M. Prasad, and S.S. Amritphale, (2019), Development and Characterization of Inorganic-Organic (Si-O-Al) Hybrid Geopolymeric Precursors via Solid State Method, Silicon, Vol. 11, pp. 221-232.
163. G. Kastiukas, S. Ruan, S. Liang, and X. Zhou, (2020), Development of Precast Geopolymer Concrete via Oven and Microwave Radiation Curing With An Environmental Assessment, Journal of Cleaner Production, Vol. 255, pp. 120290-xxx (DOI: 10.1016/j.jclepro.2020.120290)
164. J.L. Provis, G.C. Lukey, and J.S.J.Van Deventer, (xxx), Do Geopolymers Actually Contain Nanocrystalline Zeolites, A Reexamination of Existing Results, Journal details
165. S.E. Abo Sawan, M.F. Zawrah, R.M. Khattab, and A.A. Abdel-Shafi, (2019), Fabrication, Sinterabilty and Characterization of Non-Colored and Colored Geopolymers with Improved Properties, Materials Research Express, Vol. 6, pp. 075205-xxx.
166. Z. Zhang, H. Wang, and F. Mater, (2016), The Pore Characteristics of Geopolymer Foam Concrete and Their Impact on the Compressive Strength

and Modulus, Journal name, Vol. 3, pp.1-10 (https://doi.org/10.3389/fmats.2016.00038).
167. A. Kargin, V. Baev, and N. Mashkin (2017), Fly-ash Geo-Polymer Foamed Concrete, AIP Conference Proceedings 1800, Vol. 020005, pp. 020005-4 (https://doi.org/10.1063/1.4973021).
168. P. Chindaprasirt, and U. Rattanasak, (2018), Fireresistant geopolymer bricks synthesized from highcalcium fly ash with outdoor heat exposure, Clean Technologies and Environmental Policy, Vol.xxx, pp. 1-7. (https://doi.org/10.1007/s10098-018-1532-4).
169. R..Z. Sayieda, R.M. Maha, A. Doaa Ahmed, and H. Aya Mohammed, (2015), Alkali Activated Ceramic Waste with or Without two Different Calcium Sources, Advances in materials Research, Vol. 4, pp.133-144.
170. K. Komnitsas, D. Zaharaki, Vassilis Perdikatsis, (2007), Geopolymerization of Low Calcium Ferronickel Slags, Journal of Materials Science, Vol. 42, pp. 3073-3082 (DOI: 10.1007/s10853-006-0529-2).
171. J.J. Sokołowska, (2020), Long-Term Compressive Strength of Polymer Concrete-like Composites with Various Fillers, Materials, Vol. 13, pp. 1207-12 (DOI:10.3390/ma13051207).
172. S. Chandrasekhar and P.N. Pramada, (1999), Investigation on the Synthesis of Zeolite NaX from Kerala Kaolin, Journal of Porous Materials, Vol. 6, pp. 283-297.
173. M. Stachowicz and K. Granat, (2016), Influence of wet activation of used inorganic binder on cyclically refreshed water glass moulding sands hardened by microwaves, China Foundry, Vol. 13, pp. 427-432.
174. S.N. Sarmin, (2016), The Influence of Different Wood Aggregates on the Properties of Geopolymer Composites, Key Engineering Materials, Vol. 723, pp. 74-79 (DOI: 10.4028/www.scientific.net/KEM.723.74).
175. M.A. Tantawy, M.R. Shatat, A.M. El-Roudi, M.A. Taher, and M. Abd-El-Hamed, (2014), Low Temperature Synthesis of Belite Cement Based on Silica Fume and Lime, International Scholarly Research Notices, Vol. 2014, pp. 1-11.
176. C. Ferone, F. Colangelo, G. Roviello, D. Asprone, C. Menna, A. Balsamo, A. Prota, R. Cioffi, and G. Manfredi, (2013), Application-Oriented Chemical Optimization of a Metakaolin Based Geopolymer, Materials, Vol. 6, pp. 1920-1939 (DOI:10.3390/ma6051920)
177. K. Sakkas, D. Panias, P. Nomikos, and A. Sofianos, (2017), Comparison of Fire Resistant Geopolymers for Passive Fire Protection of Concrete Tunnel Linings, Open Access Library Journal, Vol. 4, pp. 1-17 (DOI: 10.4236/oalib.1103327).
178. N. Mobasher, S.A. Bernal, and John L. Provis, (2016), Structural Evolution of an Alkali Sulfate Activated Slag Cement, Journal of Nuclear Materials, Vol. 468, pp. 97-104.
179. S.-C. Ko, Alkali Activated Supersulphated Binder, WO/2000/000447 and dated: 29.06.1999, HAFFNER, Thomas, M. (Publication agent)

180. T.J. Mohammed and I.A. Saeed, (2018), Production of Geopolymer Concrete from Local Materials, Kufa Journal of Engineering, Vol. 9, pp.174-186.
181. A.R. Romero, N. Toniolo, A.R. Boccaccini, and E. Bernardo, (2019), Glass-Ceramic Foams from 'Weak Alkali Activation' and Gel-Casting of Waste Glass/Fly Ash Mixtures, Materials, Vol. 12, pp. 588-14 (DOI:10.3390/ma12040588).
182. K.E. Goetschl, M. Dietzel, B. Purgstaller, C. Grengg, and V. Mavromatis, (2021), Control of $MgSO_4$(aq) on the transformation of amorphous calcium carbonate to high-Mg calcite and long-term reactivity of the crystalline solid, Geochimica et Cosmochimica Acta, Vol. 312, pp. 357-374.
183. E.J. Rademan, A.R. Wardle, and M. Shand, Method and Compositions for Improving Performance Properties of Magnesium Oxy-chloride Cements, WO 2013/151819 Al, (International Application Number: PCT/US2013/033678, and dated: 25 March 2013), Agent: PLACKER, Jeffrey, T.; Holland & Knight LLP).
184. E.A. Azimi, M.M. Al Bakri Abdullah, P. Vizureanu, M.A.A. Mohd Salleh, A.V. Sandu, J. Chaiprapa, S. Yoriya, K. Hussin and I.H. Aziz, (2020), Strength Development and Elemental Distribution of Dolomite/Fly Ash Geopolymer Composite under Elevated Temperature, Materials, Vol. 13, pp. 1015-16 (doi:10.3390/ma13041015).
185. Jong-Kyu Lee and Jung-Sub Soh, Performance of Magnesia Cement Using MgCO3 and Serpentine, Journal of the Korean Ceramic Society Vol. 53, No. 1, pp. 116~121, 2016.
186. Davidovits, J., & Comrie, D. (1988). Archaeological long-term durability of hazardous waste disposal: preliminary results with geopolymer technologies, Division of Environmental Chemistry. American Chemical Society, Extended Abstracts, 237–240. See also: Long term durability of hazardous toxic and nuclear waste disposals. Geopolymer '88: First European Conference on Soft Mineralurgy, 125–134.
187. Comrie, D. C., Paterson, J. H., & Ritchey, D. J. (1988) Geopolymer Technologies in Toxic Waste Management. Paper presented at the Geopolymer", First European Conference on Soft Mineralurgy, Compiegne, France.
188. A. Palomoa , M.W. Grutzeckb,*, M.T. Blancoa, Alkali-activated fly ashes A cement for the future, Cement and Concrete Research 29 (1999) 1323–1329.
189. J. Davidovits, (1994), High-alkali cements for 21st century concretes. ACI Spec. Publ. SP, Vol. 144, pp. 383-398.
190. S.E. Wallah and B.V. Rangan, (2006), Fly Ash-Based Geopolymer Concrete, Research Report GC 4, Curtin University of Technology, Perth, Australia.
191. J.M. Miranda, A. Fernández-Jiménez, J.A. González, and A. Palomo, (2005), Corrosion Resistance in Activated Fly Ash Mortars, Cement and Concrete Research, Vol. 35, pp. 1210-1217 (DOI: 10.1016/j.cemconres.2004.07.030).
192. J. Davidovits, (1994), Properties of Geopolymer Cements, Conference: First International Conference Alkaline Cements and Concretes, at Kiev, Ukraine,

193. B.V. Rangan and S.E. Wallah, (2006), Low-Calcium Fly Ash-Based Geopolymer Concrete: Long-Term Properties, Research Report GC2, Curtin University of Technology, Perth, Australia.
194. A. Fernández-Jiménez, I. García-Lodeiro, and A. Palomo, (2007), Durability of Alkali-Activated Fly Ash Cementitious Materials, Journal of Materials Science, Vol. 42, pp.3055-3065, (DOI: 10.1007/s10853-006-0584-8).
195. P.K. Mehta and R.W. Burrows, (2001), Building durable structures in the 21st century, Indian Concrete Journal, Vol. 75, pp. 437-443.
196. O. Jan, W.J. Weiss and Z. Yang, (2005), Interaction between Micro-Cracking, Cracking, and Reduced Durability of Concrete: Developing Methods for Considering Cumulative Damage in Life-Cycle Modeling, Research Report FHWA/IN/JTRP-2004/10, Perdue University, West Lafayette, Indiana.
197. R. Rodrigues, S. Gaboreau, J. Gance, I. Ignatiadis, and S. Betelu, (2020), Reinforced Concrete Structures: A Review of Corrosion Mechanisms and Advances in Electrical Methods for Corrosion Monitorin, Construction and Building Materials, Elsevier, pp.121240-xxx (ff10.1016/j.conbuildmat.2020.121240ff. ffhal-02979786f).
198. E. Odd Gjorv, (2009), Durability Design of Concrete Structures in Severe Environments, Taylor & Francis, New York, NY, USA
199. S. Ahmad, (2003), Reinforcement Corrosion in Concrete Structures, Its Monitoring and Service Life Prediction–A Review, Cem. Concr. Compos. Vol. 25, pp. 459-471. (https://doi.org/10.1016/S0958-9465(02)00086-0).
200. Kasai Y, and Nakamura N (1980). Accelerated test method for durability of cement mortars in sea water. Proceedings of International Conference on Performance of Concrete in Marine Environment, St. Andrews, pp. 379-396.
201. S. Nishibayashi, K. Yamura, and S. Inoue, (1980), Durability of Concrete in Sea Water: Method of Accelerated Testing and Evaluation, International Concrete Abstracts Portal, Vol. 65, pp. 351-378.

2

Fundamentals of Geopolymer Cementitious Materials

Muktikanta Panigrahi[1]*, Ratan Indu Ganguly[2] and Radha Raman Dash[3]

[1]*PG Department of Materials Science, Maharaja Sriram Chandra Bhanja Deo University, Keonjhar Campus, Odisha, India*
[2]*Department of Metallurgical Engineering, National Institute of Technology, Rourkela, Odisha, India*
[3]*CSIR-National Metallurgical Laboratory, Jamshedpur, Jharkhand, India*

Abstract

A detailed review of the preparation of geopolymer-based materials is presented in this chapter. A literature survey is also presented to determine the progress made in geopolymer production from waste materials such as clays, silica fumes, both low-carbon and high-carbon fly ash, red mud, both high-carbon and low-carbon blast furnace slag, rice husk ash, zeolites, argon oxygen decarburization (AOD) slag, bottom ash, pond ash, etc. Different methods adopted to prepare geopolymer-based materials are given, including curing, sol-gel, deep mixing, hydrothermal synthesis and low-temperature calcination methods, mechanical activation (MA), blended mineral admixtures, an alkaline solution of thermally activated silica-alumina-bearing mineral with additives such as calcium carbonate; and an innovative *in-situ* co-reticulation process, an innovative method to reclaim the waste molding sands containing water glass with "dry" or "wet" activation of inorganic binder in waste molding sand mixtures physically hardened by microwave radiation, and sintering. The chapter also looks at the large amount of data available on the effect of variables on geopolymer production. Moreover, the formation of geopolymer structure is discussed along with the molecular structures of geopolymer materials. Finally, how the polymerization process used in condensation polymerization is similar to the process of Polycondensation polymer.

***Keywords*:** Industrial waste, pond ash, slag, geopolymer, synthesis, mechanism

**Corresponding author*: muktikanta2@gmail.com

Muktikanta Panigrahi, Ratan Indu Ganguly and Radha Raman Dash (eds.) Development of Geopolymer from Pond Ash-Thermal Power Plant Waste: Novel Constructional Materials for Civil Engineers, (71–90)

2.1 Introduction

Ordinary Portland cement (OPC) is consistently used as one of the raw materials for building worldwide. Annually, global cement production is 2.8 million tons. It is believed that the production rate is increasing by 4 million tons annually [1, 2]. Available reports indicate that there may be a need for more OPC to be produced. Therefore, the production rate is expected to sharply rise to 5.5 gigatons per year [1, 3], which will cause major environmental pollution [1–4]. It is reported that 0.54 tons of CO_2 per ton (from cement industry) and 0.46 tons of CO_2 (from other sources) are released into the atmosphere. Therefore, cement production alone can contribute approximately 7% of global emissions to our environment [1, 4–8].

Problems brought on by global warming and the greenhouse gas effect have led to several investigations being conducted to develop novel constructional binders other than cement, which will reduce CO_2 emissions [8]. The novel cementitious materials include geopolymers, which were first named and developed by Davidovits in 1978 [9]. These materials are characterized by molecular chains and network similar to organic polymers [9]. They are formed by the action of alkali molecules on inorganic compounds bearing silicon and aluminium atoms. The strength of these materials (i.e., GP) is due to chain formation as well as network formation (crosslinking like thermosetting polymer). These materials are found suitable to replace conventional ordinary Portland cement (OPC) [10]. Since they can be produced from industrial waste such as fly ash/pond ash/steel slag, the products will be very cost-effective and will also help to reduce pollution. As manufacturing of cements is an energy-intensive process, the environment is also affected due to dust released into the atmosphere, which is injurious to human health. Thus, GP can be reformed as alkali-activated cement or inorganic polymer cement. However, the development of these materials is still in its infancy. These inorganic products consist of $[SiO_4]^{-4}$ and $[AlO_4]^{-4}$ tetrahedral networks. The influence of alumino-silicate materials, such as metakaolin and low calcium fly ash, were investigated by many researchers [11, 12]. Their properties are suitable for reformation of aggregate similar to ordinary Portland cement (OPC). Based on the past results, high calcium fly ash has been successfully used as a raw material in the geopolymer mixture [13]. Current research is focused on the emerging trend of a wide range of geopolymer applications [14] in the field of new ceramic binders and matrices for hazardous waste stabilization that require fire-resistant materials and high-tech materials.

With increased use of cement in concrete, there has been an environmental threat caused by the emission of carbon dioxide during the manufacturing process. About 1.5 tons of raw materials are needed in the production of every ton of Portland cement; at the same time, about one ton of carbon dioxide (CO_2) is released into the environment during the production [1–4]. The production of PC is a costly and energy-intensive process. Additionally, global warming also occurs because of greenhouse gases such as carbon dioxide which accumulate in the atmosphere. Several studies have been carried out to reduce the use of PC in concrete for addressing global warming issues. This has created pressure on the construction industries. An attempt in this regard is the development of geopolymer concrete (GPC), which seems to be the future of construction industries. Geopolymer is an innovative binder and can be produced by treating materials of geological origin or from by-product (rich in silicon and aluminium) with highly alkaline solution [13, 14]. These include clays [15], silica fumes [16], both low-carbon [17] and high-carbon fly ash [13], red mud [18], both high-carbon [19] and low-carbon blast furnace slag [20], rice husk ash [20], zeolites [21], argon oxygen decarburization (AOD) slag [22]; and bottom ash [23], pond ash [24], man-made rock [25], biochar concrete [26], attapulgite pozzolana [27], copper mine tailings [28], activated blast furnace slag [29], low Ca electric arc ferronickel slags [30] and metakaolin [31], which are alternatives to Portland cement [1–4].

Scientists have used different methods such as curing [32], sol-gel [33], deep mixing [34], addition of additives such as Na_2CO_3, $CaCO_3$, $MgCO_3$, hydrothermal synthesis [35] and low-temperature calcination [36], mechanical activation (MA) [37], blended mineral admixtures [38]; and an alkaline solution of thermally activated silica-alumina-bearing mineral with additives such as calcium carbonate [39], innovative *in-situ* co-reticulation [40], an innovative method to reclaim waste molding sands containing water glass with "dry" or "wet" activation of inorganic binder in waste molding sand mixtures physically hardened by microwave radiation [41], sintering [42], etc. Some of the reported studies involving these methods follow.

Zheng *et al.* [33] used the sol-gel method to prepare pure Al_2O_3-$2SiO_2$ precursors (powders) for a geopolymer. The alkali-activated products derived from the precursors meet the general criteria for a geopolymer.

Vlček *et al.* [43] prepared supplementary cementitious material and studied optimization of calcined kaolin. Samples were heated at different temperatures during different times.

Wei *et al.* [44] fabricated plant fiber-based geopolymer composites using metakaolin, alkaline sodium silicate and plant fibers. Orthogonal test and single factor analysis were used to study the influence of water glass modulus, solid liquid ratio and fiber content on bending strength. The results showed the influence of factors (solid to liquid ratio, fiber content, water glass modulus) on bending strength.

Ahmari and Zhang [28] studied the use of copper mine tailings for the production of eco-friendly bricks through geopolymerization technology.

Jeon *et al.* [45] explored the preparation of activated geopolymer. They studied the beneficial effects of Na_2CO_3 as an additive for microstructural and strength properties. They analyzed $Ca(OH)_2$-activated fly ash system and NaOH-activated fly ash samples by compressive testing, XRD, SEM/BSE/EDS, 29Si/27Al MAS-NMR, MIP and TGA compared the results with the effect of Na_2CO_3.

Akhtar and Sarmah [46] used biochar, which is a carbonaceous solid material produced from three different waste sources (poultry litter, rice husk and pulp and paper mill sludge) during the preparation of concrete. They replaced cement content up to 1% of total volume. The effect of individual biochar mixed with cement on the mechanical properties of concrete was investigated through different characterization techniques.

Khudhair *et al.* [47] formulated cement and/or concrete matrix using valorized mineral and natural resources such as limestone fillers (F-Lime) and natural pozzolana (PN). The purpose of using F-Lime and PN was to minimize CO_2 emissions, energy consumption and raw materials.

Vafaei and Allahverdi [48] synthesized geopolymers by alkaline activation using waste-glass powder. They used an alkaline solution, which is a mixture of aqueous solutions of sodium hydroxide and sodium silicate having different Na_2O contents. Three types of calcium aluminate cements were also incorporated into the dry binder at levels up to 24% by weight in order to modify the chemical composition of the geopolymer source materials.

Carrasco and Puertas [49] prepared waste glass-based geopolymers using crushed residual waste glass (RWG) and activating solution which is an eco-friendly inorganic binder. The RWG was used as source of free silicon and metakaolin (MK) was taken as source of aluminosilicate.

Roviello *et al.* [50] prepared and characterized geopolymer-based hybrid composites using commercial epoxy-based organic resins (up to 25% in weight of the resin) and a metakaolin (up to 75% in weight of the resin).

Lam *et al.* [51] developed super-sulfated cement (SSC) obtained from phosphogypsum (PG) and ground granulated blast furnace slag (GFS), with a small amount of cement, which is unburnt cementitious material and environmentally friendly.

Gijbels *et al.* [52] prepared hardened binder using portlandite and ettringite intermixed with nardite and minor amounts of secondary gypsum and merwinite.

Dhavamani Doss *et al.* [53] prepared high-strength alkaline-activated concrete structures using low-calcium fly ash, ground granulated blast furnace slag (GGBS), and metakaolin, manufacturing sand (mix ratio of 1:1.31:2.22) used to build massive structures such as skyscrapers, bridges, tunnels, nuclear plants, underground structures. The water to cement ratio (for high-strength cement concrete) and the alkaline solution to binder ratio (for alkaline-activated concrete) was 0.35.

Phair *et al.* [54] established the use of aluminosilicate hydrogels as a model system for processing and developing commercial geopolymer binders.

Gong *et al.* [55] manufactured fly ash belite cement (FABC) composed of α0L-C2S and C12A7. The FABC is manufactured using low-grade fly ashes through hydrothermal process followed by low-temperature calcination.

Kalinkin *et al.* [56] synthesized geopolymers at ambient temperature using blends of fly ash and natural calcite via mechanical activation for 0–400 s in a planetary mill. The calcite content of blends was 0–10 wt%. NaOH (sodium hydroxide) solution is used as an alkaline agent.

Aboulayt *et al.* [57] prepared geopolymer in an alkaline solution using thermally activated kaolinite clay and calcium carbonate mineral (as an additive).

Rao *et al.* [58] prepared complex binders containing two or three blended mineral admixtures in terms of glass powder (GP), limestone powder (LP), and steel slag powder (SP).

Gupta *et al.* [59] developed a hybrid inorganic-organic geopolymer having Si-O-Al networks using fly ash and sodium hydroxide along with rice husk.

Kastiukas *et al.* [60] prepared geopolymer as a type of promising alternative binder. They cured the geopolymer in two ways: by oven and microwave radiation.

Provis and Bernal [61] synthesized geopolymers through alkali-silicate activation, which is a class of aluminosilicate materials cured in ambient or higher temperature. Powder X-ray diffraction, microscopy, electron diffraction, mechanical strength testing, and calorimetry were used for structure–property correlation.

Sawan *et al.* [62] prepared non-colored and colored geopolymers. Nano nickel aluminate (10 wt%) is taken as colorizing agent for colored geopolymer, whereas dried non-colored geopolymer is sintered at 800–1200 °C.

Zhang and Wang [63] manufactured geopolymer foam concretes (GFCs) in the laboratory with 0–16% additions foam.

Kargin *et al.* [64] synthesized geopolymer using fly ash. During the geopolymerization process, hydrogen peroxide (H_2O_2) solution was used as foaming agent. Activation time, and temperature of isothermal holding and hardening under normal conditions, were constant for all samples.

Somna *et al.* [65] prepared fly ash-based geopolymer which requires longer heat curing.

Zedan *et al.* [66] prepared geopolymer resin by alkali activation of ceramic waste (AACW). The AACW was replaced by 10 and 30 wt% concrete waste (CW) and 10 and 30 wt% ground granulated blast-furnace slag (GGBFS).

Komnitsas *et al.* [67] prepared geopolymer using low Ca electric arc ferronickel slags. Additives, such as kaolinite, sodium silicate, sodium hydroxide, and water, were used during the geopolymer polymerization process.

Sokołowska [68] prepared polymer composite concrete with filler and investigated durability and long-term compressive strength of polymer concrete containing common and alternative fine fillers, including quartz powder (ground sand) and by-products of the combustion of Polish fossil fuels (coal and lignite).

Chandrasekhar [69] used kaolin as cheap raw material, which is used for synthesizing zeolite. It involves a two-step reaction: dehydroxylation of kaolin (at 550–900 °C) and hydrothermal treatment of metakaolin with aqueous alkali.

Stachowicz and Granat [70] prepared an innovative binder to reclaim the waste molding sands containing water glass, which was hardened by microwave radiation. The sand mixtures consisting of high-silica sand and water-glass with average molar module 2.5, were subjected to the following cyclical process, viz., mixing the components, compacting, microwave heating, cooling-down, thermally loading the mold to 800 °C, cooling-down to ambient temperature, and knocking-out. The waste molding sands are subjected to either dry or wet activation of the binder.

Sarmin [71] prepared wood with eco-friendly binder composites for wide use in consumer products, i.e., structural and non-structural applications. They developed binder which is environmentally friendly and enhanced their sustainability. Such binder materials are known as geopolymers (i.e., alumino-silicates).

Tantawy *et al.* [72] described low temperature synthesis of belite (β-C2S) from silica fume with requisite amounts of lime and $BaCl_2$. It was treated hydrothermally in a stainless-steel capsule at 110–150 °C for 2–5 h followed by calcination at 600–700 °C for 3 h to form belite.

Ferone *et al.* [73] developed metakaolin-based geopolymeric mortar to be used as bonding matrix for external strengthening of reinforced concrete beams. During the formation of geopolymer, they varied the composition of the activating solution in terms of SiO_2/Na_2O ratio.

Sakkas *et al.* [74] developed fire-resistant geopolymer and examined its performance under thermal loading. The geopolymer was prepared using metallurgical slag, metakaolin and highly alkaline potassium hydroxide and was formed as paste. It was cured at 70 °C for a certain period of time.

Mobasher *et al.* [75] prepared the binder structure of sodium sulfate-activated slag cements. The behavior of the materials depended on the sodium sulfate content and curing duration. Ettringite, a calcium aluminium silicate hydrate (C-A-S-H) phase, and a hydrotalcite-like Mg-Al layered double hydroxide have been identified in the main reaction products. Curing had no effect on the products. Upon curing at advanced ages, the most significant changes were observed on strength properties.

Suz-Chung Ko [76] described an activated supersulphated aluminosilicate binder containing aluminosilicate, calcium sulphate and an activator containing alkali metal salts. The selected group (i.e., aluminosilicate) was obtained from blast furnace slag, clay, marl and industrial by-products (such as fly ash).

Mohammed and Saeed [77] produced geopolymer concrete, which is a new construction material in Iraq. Foreign fly ash and local fly ash obtained from Baghdad South Gas Power Plant were used with good specifications to obtain geopolymer concrete. In addition, the local pozzolana material was used to explore the possibility of production of geopolymer concrete.

Rincón *et al.* [78] produced glass foams formed by alkali activation and gel casting. The alkali activation is done by soda-lime waste glass powders that is allowed to obtain a well-dispersed concentrated suspension. It underwent gelification by curing at low temperature (75 °C).

Purgstaller *et al.* [79] prepared temporal amorphous stabilizer using carbonate mineral. It was prepared either in biotic environments or in abiotic environments.

Rademan *et al.* [80] prepared magnesium oxy chloride (sorel)-based cements and magnesium oxy sulfate-based cements, which were water and corrosion resistance. They have incorporated various alkali metal phosphates, such as $MgHPO_4$ or MgH_2PO_4, with alkali metal fatty acids (magnesium stearate; and metal or alkali metal sulfates such as aluminum sulfate or magnesium sulfate). Moreover, water resistance is enhanced by either pre-carbonating the mix water or the liquid magnesium chloride phase of the cements, or by adding a carbonate into the powder phase.

Accelerated curing of this system was also obtained by using various inorganic metal oxides.

Azimi *et al.* [81] used fly ash and dolomite as raw materials during the production of geopolymer composite using alkaline solution (sodium hydroxide, and liquid sodium silicate).

Lee and Soh [82] developed carbon-negative cement. Two of the methods most typically used for reducing CO_2 in building materials were either addition of slag and fly ash, like pozzolana material, or developing carbon-negative cement.

2.2 Parameters of Geopolymer Concrete

Effects of some parameters on the compressive strength of low-calcium fly ash-based geopolymer concrete are discussed below. Parameters considered are as follows:

1) Ratio of alkaline liquid-to-fly ash, by mass
2) Molar concentration of sodium hydroxide (NaOH) in solution
3) Ratio of sodium silicate solution-to-sodium hydroxide solution
4) Curing temperature
5) Curing time
6) Handling time
7) Addition of superplasticizer
8) Rest period prior to curing
9) Water content of mixture
10) Dry curing or steam curing
11) Mixing time
12) Age of concrete

2.3 Geopolymer Formation Mechanism

Geopolymers are known as inorganic polymers having covalently bonded alumino-silicate chains. They are noncrystalline, i.e., amorphous in nature [83]. Repeating units of geopolymer are basically a mineral compound or mixture of compounds such as silico-oxide (-Si-O-Si-O-), silico-aluminate (-Si-O-Al-O-), ferro-silico-aluminate (-Fe-O-Si-O-Al-O-) or alumino-phosphate (-Al-O-P-O-) [84]. Repeating units are formed by the process of polymerization, called the geopolymerization process. One of the polymerization routes is treating aluminosilicate compounds with alkali solution containing Na^+, K^+, Li^+, Ca^{2+}, Cs^+ and the like.

Geopolymer framework structure is produced by condensation polymerization process. The monomeric unit of geopolymer consists of tetrahedral aluminosilicate units with alkali metal as an ion, balancing charge associated with tetrahedral Al. Geopolymerization occurs at either ambient or slightly elevated temperature. Leaching of solid aluminosilicate raw materials in alkaline solutions leads to the transfer of leached species from solid surfaces into a growing gel phase. It is followed by nucleation and condensation of gel phase [85–89].

The fundamental unit consists of a small cation, such as Si^{+4}, or Al^{+3}, tetrahedrally coordinated with four oxygens. Many reports have explained the geometry of $SiO4^{+4}$ tetrahedron and other mineral structures [85–89].

Ionic tetrahedral coordination is no longer adapted to explain geopolymer chemistry. Two tetrahedrons are shared by one oxygen anion O^{2-}, i.e., Si-O-Si- structure. A covalent bond is attained between the Si and O atom. Both atoms are shared, one electron each. This leads to the formation of stronger bond within the structure. Gibbs *et al.* [85–89] studied the polymeric bond, i.e., Si-O-Si-O. They successfully modeled properties and structures of silicate tetrahedron units. This offers credence to a silica polymorph, i.e., giant molecule bound together, which is similar to quartz structure, i.e., Si-O-Si skeleton. It is a small siloxane molecule called an oligomer [85–91]. Siloxane oligomers have the same structure as the silico-aluminate oligomers [85–91].

Geopolymerization process is carried out with pond ash and alkali activators (NaOH and Na_2SiO_3). During the process, many small molecules are formed, which are known as oligomers (dimer, trimer, tetramer, pentamer). Due to covalent bonding of reactants, three-dimensional (3D) networks are resulted. North and Swaddle [92] ascertained the existence of soluble isolated alumino-silicate molecules with high concentrations, which were stable at higher temperatures as well as in an environment of high pH. Schematic diagrams of oligomers are shown in Figure 2.1.

The -Si-O-Al-O- skeleton formed during the geopolymerization process occurs due to the reaction of metakaolin MK-750 with alkali solution [93]. The skeleton encompasses four main phases (Figure 2.1) such as alkaline depolymerization of the poly(siloxo) layer of kaolinite, formation of monomeric and oligomeric species (ortho-sialate" $(OH)_3$-Si-O-Al-$(OH)_3$ molecule), soluble K-polysiloxonate, ortho-sialate-disiloxo cyclic structure (Figure 2.1). The hydroxide is liberated by condensation reactions. It may react again and again to form higher oligomers and polymeric

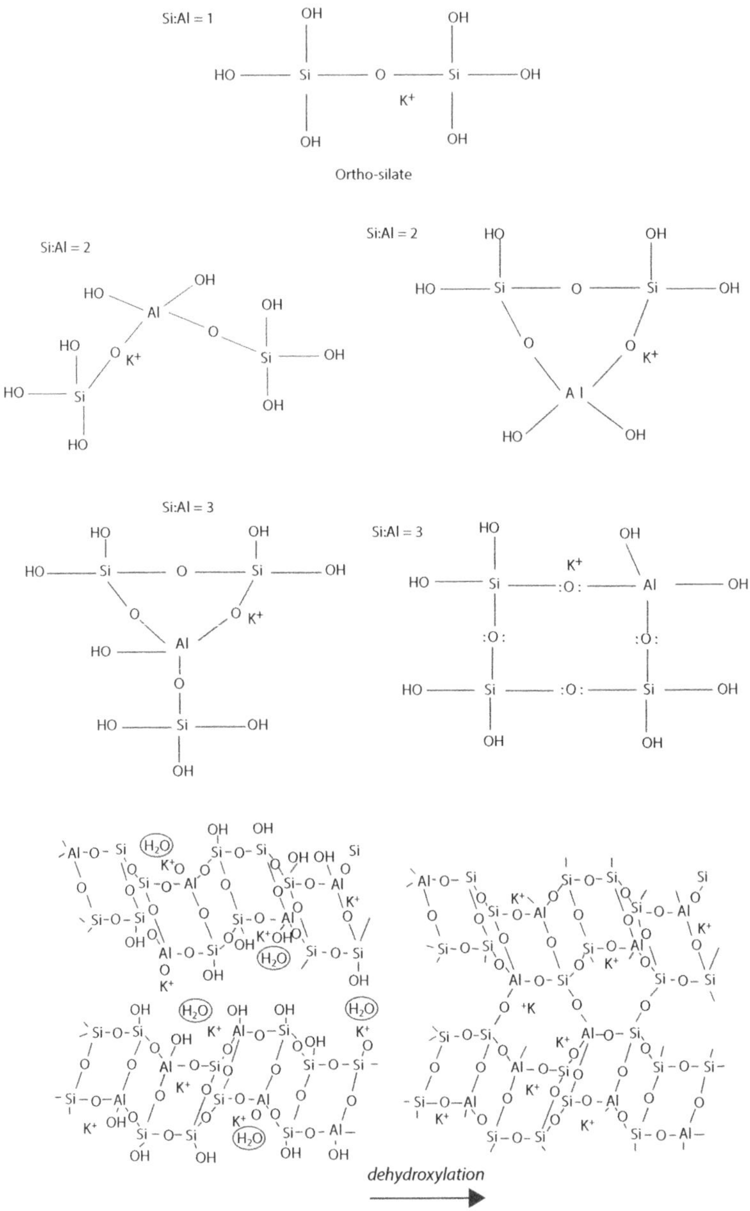

Figure 2.1 Polycondensation into zeolite-like 3D frameworks [91].

3D-networks. A similar skeleton (Si-O-Al-O-) is also observed in geopolymer by fly ash in alkaline medium [94].

The skeleton involves five main phases, viz., nucleation stage (aluminosilicate formation from fly ash particle, dissolution in alkaline medium), releasing aluminates and silicates (as monomers), dimers (joining of monomers to form), and further interlinking with other monomers to form trimers, tetramers and so on. When the solution reaches saturation, an aluminum-rich gel (Gel 1) precipitates. As the reaction progresses, more Si-O groups from the initial solid source dissolve, increasing the silicon concentration in the medium and gradually raising the proportion of silicon in the zeolite precursor gel (Gel 2).

The skeleton of geopolymer is mainly aluminosilicate frameworks, which is similar to rock-forming minerals. Some related reports are available on the theoretical structure. In 1994, Davidovits [92] showed a theoretical structure for K-poly(sialate-siloxo) (K)-(Si-O-Al-O-Si-O), which is consistent with the NMR spectrum.

2.4 Conclusions

Basic information on geopolymer cementitious materials was provided in this chapter. Literature relevant to the fly ash-based geopolymer was presented in brief. More reviews on the specific properties are presented in the relevant chapters of this book. The problems and challenges regarding fly ash-based geopolymer have been identified. Since a significant gap of research in the area of geopolymer concrete for ambient curing condition exists, the research goal is to eliminate the necessity of heat curing for fly ash-based geopolymer. Therefore, mixtures modified with additives and the methodology of experimental works were also detailed in this chapter.

Acknowledgments

The authors convey their sincere thanks to the Ministry of Mines, Government of India, for providing the financial grant to carry out this work [Grant number= F.No.:14/54/20214-Met-IV dated: 29.12.2014]. They also convey their sincere thanks to GIET, Gunupur, Rayagada, Odisha, India, for providing lab facilities to do the research work. The authors would also like to thank the CRF, IIT Kharagpur, for providing their testing facilities for developed Geopolymer.

References

1. M. Schneider, M. Romer, M. Tschudin, and H. Bolio, (2011), Sustainable Cement Production-Present and Future, Cem. Concr. Res. Vol. 41, pp. 642-650, (http://dx.doi.org/10.1016/j.cemconres.2011.03.019).
2. D.A. Salas, A.D. Ramirez, C.R. Rodríguez, D.M. Petroche, A.J. Boero, and J. Duque-Rivera, (2016), Environmental Impacts, Life Cycle Assessment and Potential Improvement Measures for Cement Production: A Literature Review, J. Clean. Prod. Vol. 113, pp. 114-122 (http://dx.doi.org/10.1016/j.jclepro.2015.11.078).
3. M. Babaee and A. Castel, (2016), Chloride-Induced Corrosion of Reinforcement in Low Calcium Fly Ash-based Geopolymer Concrete, Cem. Concr. Res., Vol. 88, pp. 96-107 (http://dx.doi.org/10.1016/j.cemconres.2016.05.012).
4. B.W. Jo, J.S. Choi, K.W. Yoon, and J.H. Park, (2012), Material Characteristics of Zeolite Cement Mortar, Constr. Build. Mater. Vol. 36, pp. 1059-1065 (http://dx.doi.org/10.1016/j.conbuildmat.2012.06.061).
5. C. Gunasekara, D.W. Law, S. Setunge, and J.G. Sanjayan, (2015), Zeta Potential, Gel Formation and Compressive Strength of Low Calcium Fly Ash Geopolymers, Constr. Build. Mater. Vol. 95, pp. 592-599 (http://dx.doi.org/10.1016/j.conbuildmat.2015.07.175).
6. P. Duan, C. Yan, W. Zhou, and W. Luo, (2016), Fresh Properties, Mechanical Strength and Microstructure of Fly Ash Geopolymer Paste Reinforced with Sawdust, Constr. Build. Mater. Vol. 111, pp. 600-610 (http://dx.doi.org/10.1016/j.conbuildmat.2016.02.091).
7. A. Petrillo, R. Cioffi, C. Ferone, F. Colangelo, and C. Borrelli, (2016), Eco-Sustainable Geopolymer Concrete Blocks Production Process, Agric. Agric. Sci. Procedia, Vol. 8, pp. 408-418 (http://dx.doi.org/10.1016/j.aaspro.2016.02.037).
8. K.H. Yang, J.K. Song, and K. Il Song, (2013), Assessment of CO_2 Reduction of Alkali-Activated Concrete, J. Clean. Prod., Vol. 39, pp. 265-272 (http://dx.doi.org/10.1016/j.jclepro. 2012.08.001).
9. J. Davidovits, (1991) Geopolymers, J. Therm. Anal. Vol. 37, pp. 1633-1656 (http://dx.doi.org/10.1007/BF01912193).
10. K.F. Tee and S. Mostofizadeh, (2021), Review A Mini Review on Properties of Portland Cement Concrete with Geopolymer Materials as Partial or Entire Replacement, Infrastructures, Vol. 6, pp. 1-21 (https://doi.org/10.3390/infrastructures6020026).
11. C. Monticelli, M.E. Natali, A. Balbo, C. Chiavari, F. Zanotto, S. Manzi, and M.C. Bignozzi, (2016), A Study on the Corrosion of Reinforcing Bars in Alkali-Activated Fly Ash Mortars Under Wet and Dry Exposures to Chloride Solutions, Cem. Concr. Res., Vol. 87, pp. 53-63 (http://dx.doi.org/10.1016/j.cemconres.2016.05.010).

12. Authors name (2017), Geopolymers: Ceramic-like Inorganic Polymers, Journal of Ceramic Science and Technology, Vol. 8, pp. 335-350 (DOI: 10.4416/JCST2017-00038)
13. G. Lavanya and J. Jegan (2015), Durability Study on High Calcium Fly Ash Based Geopolymer Concrete, Advances in Materials Science and Engineering Vol. 2015, pp.1-7 (http://dx.doi.org/10.1155/2015/731056, pp. 1-8).
14. K. Srinivasan and A. Sivakumar, (2013), Review Article Geopolymer Binders: A Need for Future Concrete Construction, ISRN Polymer Science, Vol. 2013, pp.1-8 (http://dx.doi.org/10.1155/2013/509185).
15. N. HamdiabImen, B. Messaouda, and E. Srasraa, (2019), Production of geopolymer binders using clay minerals and industrial wastes, Comptes Rendus Chimie, Vol. 22, pp. 220-226.
16 O.H. Li, L. Yun-Ming, H. Cheng-Yong, R. Bayuaji, M.M. Al Bakri Abdullah, F.K. Loong, T.S. Jin, N.H. Teng, M. Nabiałek, B. Jez and N.Y. Sing, (2021), Evaluation of the Effect of Silica Fume on Amorphous Fly Ash Geopolymers Exposed to Elevated Temperature, Magnetochemistry, Vol. 7, pp.1-14 (https://doi.org/10.3390/magnetochemistry7010009).
17. T. Sathanandam, P.O..Awoyera, V. Vijayan, and K. Sathishkumar, (2017), Low Carbon Building: Experimental Insight on the Use of Fly Ash and Glass Fibre for Making Geopolymer Concrete, Sustainable Environment Research, Vol. 27, pp. 146-153.
18. Y. Li, X. Min, Y. Ke, D. Liu, and C. Tang, (2019), Preparation of Red Mud-Based Geopolymer Materials from Mswi Fly Ash and Red Mud by Mechanical Activation, Waste Management, Vol. 83, pp. 202-208.
19. M. Panigrahi, R.I. Ganguly, and R.R. Dash, (2018), An Advanced Cured High Carbon Ferrochrome Slag (HCFCS) Geopolymer (GP): A Constructional Materials, IOP Conference Series: Materials Science and Engineering, Vol.410, pp.1-15.
20. G. Kürklü, (2016), The Effect of High Temperature on the Design of Blast Furnace Slag and Coarse Fly Ash-Based Geopolymer Mortar, Composites Part B: Engineering, Vol. 92, pp. 9-18.
21. C.Villa, E.T.Pecina, R.Torres, and L.Gómez, (2010), Geopolymer Synthesis using Alkaline Activation of Natural Zeolite, Construction and Building Materials, Vol. 24, pp. 2084-2090.
22. J.P. Singh D.K. Sinha, S. Kumar, M.K. Jain, A. Kumar, (2018), ETP Sludge as Filler and the Role of AOD Slag and GBFS in Fly Ash–Slag–Sludge Blended Geopolymer, MOJ Civil Eng., Vol. 4, Vol. 379-384 (DOI: 10.15406/mojce.2018.04.00132).
23. L.J.J. aramillo, N.F. Elyseu, S.G. Marianade, S.P. Erick, Z.V. Adriano, and M. Bernardin, (2021), Use of Fly and Bottom Ashes From a Thermoelectrical Plant in the Synthesis of Geopolymers: Evaluation of Reaction Efficiency, Energy Geoscience, Vol. 2, pp. 167-173.
24. M. Panigrahi, R.R. Dash, and R.I. Ganguly, (2018), Development of Novel Constructional Material from Industrial Solid Waste as Geopolymer for

Future Engineers, IOP Conference Series: Materials Science and Engineering, Vol. 410, pp.1-12.

25. J. Davidovits, (1994), Geopolymers: Man-Made Rock Geosynthesis and the Resulting Development of Very Early High Strength Cement, Journal of Materials Education, Vol.16, pp 91-139.
26. A.K Sarmah, (2017), Novel Biochar-Concrete Composites: Manufacturing, Characterization and Evaluation of the Mechanical Properties, Science of The Total Environment, Vol. 616, pp. 408-416 (DOI: 10.1016/j.scitotenv.2017.10.319).
27. Mouhamadou Bassir Diop, Laurent Molez, Ahmed Bouguerra, Arona Ndofféne Diouf, Michael W. Grutzeck, (2014), Manufacturing Brick from Attapulgite Clay at Low Temperature by Geopolymerization, Arabian Journal for Science Engineering, Vol. 39, pp. 4351-4361 (DOI 10.1007/s13369-014-1007-9).
28. S. Ahmari and L. Zhang, (2012), Production of Eco-Friendly Bricks from Copper Mine Tailings Through Geopolymerization, Construction and Building Materials, Vol. 29, pp. 323-331.
29. X. Huang, T. Huang, S. Li, F. Muhammad, G. Xu, Z. Zhao, L. Yu, Y. Yan, D. Li, and B. Jiao, (2016), Immobilization of Chromite Ore Processing Residue with Alkali-Activated Blast Furnace Slag-based Geopolymer, Ceramics International, Vol. 42, pp. 9538-9549.
30. K. Komnitsas, D. Zaharaki, and V. Perdikatsis, (2007), Geopolymerization of Low Calcium Ferronickel Slags, Journal of Materials Science, Vol. 42, pp. 3073-3082 (DOI: 10.1007/s10853-006-0529-2).
31. L. Chen, Z. Wang, Y. Wang, and J. Feng, (2016), Preparation and Properties of Alkali Activated Metakaolin-Based Geopolymer, Materials, Vol. 9, pp. 767-xxx (DOI:10.3390/ma9090767).
32. B. Nematollahi, J. Sanjayan, F. Uddin, and A. Shaikh, (2015), Synthesis of Heat and Ambient Cured One-Part Geopolymer Mixes with Different Grades of Sodium Silicate, Ceramics International, Vol. 41, pp. 5696-5704.
33. G. Zheng, X. Cui, W. Zhang, and Z. Tong, (2009), Preparation of Geopolymer Precursors by Sol–Gel Method and Their Characterization, Journal of Materials Science, Vol. 44, pp. 3991–3996.
34. H. Canakci, H. Güllü, and A. Alhashemy, (2019), Performances of Using Geopolymers Made with Various Stabilizers for Deep Mixing, Materials, Vol. 12, 2542-32; (DOI:10.3390/ma12162542).
35. N. K. Lee, Hammad R. Khalid, H.K. Lee, (2016), Synthesis of Mesoporous Geopolymers Containing Zeolite Phases by a Hydrothermal Treatment, Microporous and Mesoporous Materials, Vol. 229, pp. 22-30 (DOI: 10.1016/j.micromeso.2016.04.016)
36. N.H. Jamil, M.M. Al Bakri Abdullah, F. Che Pa, H. Mohamad, W.M.A.W. Ibrahim, P. Amonpattaratkit, J. Gondro, W. Sochacki and N. Ibrahim, (2021), Self-Fluxing Mechanism in Geopolymerization for Low-Sintering

Temperature of Ceramic, Materials, Vol. 14, pp. 1325-12 (https://doi.org/10.3390/ma14061325).
37. B. Wei, Y. Zhang, and S. Bao, (2017), Preparation of Geopolymers from Vanadium Tailings by Mechanical Activation, Construction and Building Materials, Vol. 145, pp. 236-242 (DOI: 10.1016/j.conbuildmat.2017.03.234).
38. S Md Khalid, M.S. Shobha, H.M. Tanu, T.V. Reshma, (2021), Ternary Blended Geo-Polymer Concrete–A Review, IOP Conf. Series: Earth and Environmental Science, Vol. 822, pp. 012043 (DOI:10.1088/1755-1315/822/1/012043).
39. P. Cong and Y. Cheng, (2021), Advances in Geopolymer Materials: A Comprehensive Review, Journal of Traffic and Transportation Engineering (English Edition), Vol. 8, pp. 283-314.
40. F. Colangelo, G. Roviello, L. Ricciotti, C. Ferone, and R, Cioffi, (2013), Preparation and Characterization of New Geopolymer-Epoxy Resin Hybrid Mortars, Materials (Basel), Vol. 6, pp. 2989-3006.
41. W. Jina, F. Zitian, Z. Xiaolei, and P. Di, (2009), Propertis of sodium silicate bonded sand hardened by microwave heating, China Foundry, Vol. 6, pp. 191-196.
42. R. Ahmad, M.M. Al Bakri Abdullah, W.M.W. Ibrahim, K. Hussin , F.H. Ahmad Zaidi , J. Chaiprapa, J.J. Wysłocki, K.B. łoch and M. Nabiałek, (2021), Role of Sintering Temperature in Production of Nepheline Ceramics-Based Geopolymer with Addition of Ultra-High Molecular Weight Polyethylene, Materials, Vol. 14, pp. 1077-xxx (https://doi.org/10.3390/ma14051077).
43. J. Vlček, M. Topinková, M. Klárová, P. Maierová, H. Ovčačíková, V. Matějka, A. Martaus, and V. Blahůšková, (2021), Alkali-Activated Metakaolin and Fly Ash as Unfired Ceramic Bonding Systems, Minerals, Vol. 11, pp. 197-xxx (https://doi.org/10.3390/min11020197).
44. S.J. Wei, J.L. Tan, W.L. Lu, L.P. Liu, S.J. Yu, and G.J. Zheng, (2018), Preparation and Performances of Geopolymer-Based Plant Fiber Composites, Solid State Phenomena, Vol. 281, pp. 266-271.
45. D. Jeon, Y. Jun, Y. Jeong, and J.E. Oh, (2015), Microstructural and Strength Improvements through the Use of Na_2CO_3 in a Cementless $Ca(OH)_2$-Activated Class F Fly Ash System, Cement and Concrete Research, Vol. 67, pp. 215-225 (DOI: 10.1016/j.cemconres.2014.10.001).
46. A. Akhtar and A.K. Sarmah, (2017), Novel Biochar-Concrete Composites: Manufacturing, Characterization and Evaluation of the Mechanical Properties, Science of the Total Environment, Vol. 616-617, pp. 408-416. (doi: 10.1016/j.scitotenv.2017.10.319).
47. M. Khudhair, M.S. Elyoubi, A. Essamri, and A. Elharfi, (2018), Characterization and Formulation of a New Eco-Friendly Hydraulic Binder based on Combination of Inorganic and Organic Admixtures, MATEC Web of Conferences 149:01069, (DOI: 10.1051/matecconf/201814901069).
48. M. Vafaei and A. Allahverdi, (2017), High Strength Geopolymer Binder based on Waste-Glass Powder, Advanced Powder Technology, Vol. 28, pp. 215-222.

49. M.T. Carrasco and F. Puertas, (2014), Waste Glass in the Geopolymer Preparation. Mechanical and microstructural characterization, Journal of Cleaner Production, Vol. 90, pp. xxx-xxx (DOI: 10.1016/j.jclepro.2014.11.074).
50. G. Roviello, C. Menna, O. Tarallo, L. Ricciotti, F. Messina, C. Ferone, D. Asprone, R. Cioffi, (2017), Lightweight geopolymer-based hybrid materials, Composites Part B Engineering, Vol. 128, (DOI: 10.1016/j.compositesb.2017.07.020).
51. N.N. Lam, (2018), A study on super-sulfated cement using Dinh Vu phosphogypsum, IOP Conference Series Earth and Environmental Science, Vol. 143, pp. 012016-xxx (DOI: 10.1088/1755-1315/143/1/012016).
52. K. Gijbels, H. Nguyen, P. Kinnunen, W. Schroeyers, Y. Pontikes, S. Schreurs, M. Illikainen, (2019), Feasibility of Incorporating Phosphogypsum in Ettringite-based Binder from Ladle Slag, Journal of Cleaner Production, Vol. 237, pp. 117793-10.
53. S. Dhavamani Doss, S. Thirugnanasambandam, P. Murthi, K. Poongodi, (2020), Development of Alkaline Activated High Strength Concrete using Fly Ash-Ground Granulated Blast Furnace Slag-Metakaolin as Binders and Manufacturing Sand as Fine Aggregate, International Journal of Innovative Technology and Exploring Engineering (IJITEE) Vol 9, pp. xxx-x.
54. J. Phair, J.D. Smith, and J.S.J. Van Deventer, (2003), Characteristics of Aluminosilicate Hydrogels Related to Commercial "Geopolymers, Materials Letters, Vol. 57, pp. 4356-4367 (DOI: 10.1016/S0167-577X(03)00325-2).
55. Y. Gong, J. Yang, H. Sun, and F. Xu, (2021), Effect of Fly Ash Belite Cement on Hydration Performance of Portland Cement, Crystals, Vol. 11, pp. 740-xxx (https://doi.org/10.3390/ cryst11070740).
56. A.M. Kalinkin, B.I. Gurevich, E.V. Kalinkina, M.V. Chislov, I.A. Zvereva, (2021), Geopolymers Based on Mechanically Activated Fly Ash Blended with Dolomite, Minerals, Vol. 11, pp.700-xxx (DOI: 10.3390/min11070700).
57. A. Aboulayt, M. Riahi, M.O. Touhami, H. Hannache, M. Gomina, and R. Moussa, (2017), Properties of Metakaolin based Geopolymer Incorporating Calcium Carbonate, Advanced Powder Technology, Vol. 28, pp. xxx-xxx (DOI: 10.1016/j.apt.2017.06.022).
58. M. Rao, J. Wei, Z. Gao, W. Zhou, Q. Li, and S. Liu, (2016), Study on Strength and Microstructure of Cement-Based Materials Containing Combination Mineral Admixtures, Advances in Materials Science and Engineering, Vol. 2016, pp. 1-10 (http://dx.doi.org/10.1155/2016/7243670).
59. R. Gupta, P. Bhardwaj, K. Deshmukh, D. Mishra, M. Prasad, and S.S. Amritphale, (2019), Development and Characterization of Inorganic-Organic (Si-O-Al) Hybrid Geopolymeric Precursors via Solid State Method, Silicon, Vol. 11, pp. 221–232.
60. G. Kastiukas, S. Ruan, S. Liang, X. Zhou, (2020), Development of precast geopolymer concrete via oven and microwave radiation curing with an environmental assessment, Journal of Cleaner Production, Vol. 255, pp. 120290-xxx (DOI: 10.1016/j.jclepro.2020.120290).

61. J.L. Provis and S.A. Bernal, (2014). Geopolymers and Related Alkali-Activated Materials. Annu. Rev. Mater. Res., Vol. 44, pp. 299-327 (https://doi.org/10.1146/annurev-matsci-070813-113515).
62. S.E. Abo Sawan, M.F. Zawrah, R.M. Khattab, and A.A. Abdel-Shafi, (xxx), Fabrication, Sinterabilty and Characterization of Non-colored and Colored Geopolymers with Improved Properties, Materials Research Express, Vol. 6, pp.xxxx, (DOI: 10.1088/2053-1591/ab1454).
63. Z. Zhang and H. Wang, (2016), The Pore Characteristics of Geopolymer Foam Concrete and Their Impact on the Compressive Strength and Modulus, Front. Mater., Vol. 3, pp.1-10 (https://doi.org/10.3389/fmats.2016.00038).
64. A. Kargin, V. Baev, and N. Mashkin, Fly-Ash Geo-Polymer Foamed Concrete, AIP Conf. Proc. 1800, 020005-1–020005-4, ISBN: 978-0-7354-1468-6/$30.00 020005-5 (DOI: 10.1063/1.4973021)
65. K. Somna and C. Jaturapitakkul, P. Kajitvichyanukul, and P. Chindaprasirt, (2011), NaOH-activated ground fly ash geopolymer cured at ambient temperature, Fuel, Vol. 90, pp. 2118-2124 (DOI: 10.1016/j.fuel.2011.01.018).
66. R. Zedan, Sayieda, R. Mohamed, Maha, A. Ahmed, D. Mohammed, H. Aya, (2015), Alkali Activated Ceramic Waste with or without two Different Calcium Sources, Advances in Materials Research, Vol. 4, pp.133-144.
67. K. Komnitsas, D. Zaharaki, and V. Perdikatsis, (2007), Geopolymerisation of Low Calcium Ferronickel Slags, Journal of Materials Science, Vol. 42, pp. 3073-3082 (DOI: 10.1007/s10853-006-0529-2).
68. J.J. Sokołowska, (2020), Long-Term Compressive Strength of Polymer Concrete-like Composites with Various Fillers, Materials, Vol. 13, pp. 1207-xxx (DOI:10.3390/ma13051207).
69. S. Chandrasekhar, (1996) Influence of Metakaolinization Temperature on the Formation of Zeolite 4A from Kaolin, Clay Minerals, Vol. 31, pp. 253-261.
70. M. Stachowicz and K. Granat, (2016) Influence of Wet Activation of Used Inorganic Binder on Cyclically Refreshed Water Glass Moulding Sands Hardened By Microwaves, China Foundry, Vol. 13, pp. 427-432.
71. S.N. Sarmın, (2014), Investigating the Possibility of Geopolymer to Produce Inorganic-Bonded Wood Composites for Multifunctional Construction Material–A Review, Bioresources, Vol. 9, pp. 7941-7950.
72. M.A. Tantawy, M.R. Abdalla Shatat, A.M. El-Roudi, M.A. Taher, and M. Abd-El-Hamed, (2014), (2014), Low Temperature Synthesis of Belite Cement Based on Silica Fume and Lime, International Scholarly Research Notices, Vol. 2014 (DOI: 10.1155/2014/873215).
73. C. Ferone, F. Colangelo, G. Roviello, D. Asprone, C. Menna, A. Balsamo, A. Prota, R. Cioffi, and G. Manfredi, (2013), Application-Oriented Chemical Optimization of a Metakaolin Based Geopolymer, Materials, Vol. 6, pp. 1920-1939 (https://doi.org/10.3390/ma6051920).
74. K. Sakkas, S. Kapelari, D. Panias, P. Nomikos, and A. Sofianos, (2014), Fire Resistant K-Based Metakaolin Geopolymer for Passive Fire Protection of

Concrete Tunnel Linings, Open Access Library Journal, Vol. 1, pp. 1-10 (DOI:10.4236/oalib.1100806).

75. N. Mobasher, S.A. Bernal and J.L. Provis, (2016), Structural evolution of an alkali sulfate activated slag cement, Journal of Nuclear Materials, Vol. 468, pp.97-104 (http://dx.doi.org/10.1016/j.jnucmat.2015.11.016).
76. S.-C. Ko, WO2000000448A1 and dated 6 Jan 2000, Activated aluminosilicate binder,
77. Thaer J. Mohammed, and Ikram A. Saeed, (2018), Production of Geopolymer Concrete from Local Materials, Kufa Journal of Engineering Vol. 9, pp.174-186.
78. A. Rincón, G. Giacomello· M. Pasetto, and E. Bernardo, (2017), Novel 'Inorganic Gel Casting' Process for the Manufacturing of Glass Foams, Journal of the European Ceramic Society, Vol.37, pp. 2227-2234.
79. B. Purgstaller, K.E. Goetschl, V. Mavromatis, and M. Dietzel, (2019), Solubility investigations in the Amorphous Calcium Magnesium Carbonate System, Cryst. Eng. Comm, Vol. 21, pp. 155-164 (DOI: 10.1039/C8CE01596A).
80. J.E. Rademan, R.Wardle, M. Shand, (2013), Method and Compositions for Improving Performance Properties of Magnesium Oxychloride Cements, Patent application number: WO2013151819A1 and dated: 25 March 2013 (25.03.2013) (Agent: PLACKER, Jeffrey, T.; Holland & Knight LL 10 St. James Avenue, Boston, Massachusetts 021 16, US).
81. E.A. Azimi, M.M. Al Bakri Abdullah, P. Vizureanu, M.A.A. Mohd Salleh, A.V. Sandu, J. Chaiprapa, S. Yoriya, H. Kamarudin, and I.H.A. Aziz, (2020), Strength Development and Elemental Distribution of Dolomite/Fly Ash Geopolymer Composite under Elevated Temperature, Materials, Vol. 13, pp.1015-xxx (DOI: 10.3390/ma13041015).
82. J.-K. Lee and J.-S. Soh, (2016), Performance of Magnesia Cement Using $MgCO_3$ and Serpentine, Journal of the Korean Ceramic Society, Vol. 53, pp. 116-121 (DOI: 10.4191/kcers.2016.53.1.116).
83. N.I. Kozhukova, R.V. Chizhov, I.V. Zhervovsky, and V.V. Strokova, (2016). Structure Formation of Geopolymer Perlite Binder Vs. Type of Alkali Activating Agent, International Journal of Pharmacy & Technology, Vol. 8, pp. 15339-xxx.
84. "What is a geopolymer? Introduction-Geopolymer Institute, (https://www.geopolymer.org/science/introduction)
85. Y. Huang, and M. Han, (2011), The influence of α-Al_2O_3 Addition on Microstructure, Mechanical and Formaldehyde Adsorption Properties of Fly Ash-Based Geopolymer Products, Journal of Hazardous Materials, Vol. 193, pp. 90-94.
86. K. Pimraksaa, P. Chindaprasirt, A. Rungchet, K. Sagoe-Crentsil, and T. Sato, (2011), Lightweight Geopolymer Made of Highly Porous Siliceous Materials with various Na_2O/Al_2O_3 and SiO_2/Al_2O_3 ratios, Materials Science and Engineering A, Vol. 528, pp. 6616-6623.

87. D. Feng, J.L. Provis, V. Deventer, and S.J. Jannie (2012), Thermal Activation of Albite for the Synthesis of One-Part Mix Geopolymers, J. Am. Ceram. Soc., Vol. 95, pp.565-572.
88. P. He, D. Jia, M. Wang, Y. Zhou, (2011), Thermal Evolution and Crystallization Kinetics of Potassium-based Geopolymer, Ceramics International, Vol. 37, pp. 59-63.
89. L.J. Bell, E.P. Driemeyer, and M.W. Kriven, (2009), Formation of Ceramics from Metakaolin-Based Geopolymers. Part II: K-Based Geopolymer, J. Am. Ceram. Soc., Vol. 92, pp. 607-615.
90. https://www.geopolymer.org/science/about-geopolymerization
91. G.V. Gibbs, F.C. Hill, M.B. Jr, Boisen and R.T. Downs, (2000), Molecules as a Basis for Modeling the Force Field of Silica, Chapter 6 in Structure and Imperfections in Amorphous and Crystalline Silicon Dioxide, Edited by R. A. B. Devine, J.-P. Duraud and E. Dooryhee, John Wiley & Sons Ltd
92. M.R. North, and T.W. Swaddle, (2000), Kinetics of Silicate Exchange in Alkaline Aluminosilicate Solutions, Inorg. Chem., Vol. 39, pp. 2661-2665.
93. https://www.geopolymer.org/science/about-geopolymerization
94. P. Duxson, A. Fernández-Jiménez, J.l. Provis, G.C. Lukey, A. Palomo, and J.S.J. Van Deventer, (2007), Geopolymer Technology: the Current State of the Art, J. Mat. Sci., Vol. 42, pp. 2917-2933.

3

Pond Ash (PA)-Based Geopolymer Cementitious Materials

Muktikanta Panigrahi[1]*, Ratan Indu Ganguly[2] and Radha Raman Dash[3]

[1]PG Department of Materials Science, Maharaja Sriram Chandra Bhanja Deo University, Keonjhar Campus, Odisha, India
[2]Department of Metallurgical Engineering, National Institute of Technology, Raurkela, Odisha, India
[3]CSIR-National Metallurgical Laboratory, Jamshedpur, Jharkhand, India

Abstract
The focus of this chapter is the use of pond ash as raw material in geopolymer cements. The process parameters of the study were defined after review of the literature. The chemicals needed for geopolymerization were collected; and the pond ash was procured from Damanjodi, Odisha (M/S NALCO). The pond ash materials were dried in the open air (1 day) followed by drying in an oven at 120 °C for 12 h. They were classified by sieve analysis and mixed in certain proportions, as descried in the study. They were subsequently mixed manually as well as mechanically, and were then dried in an oven for a certain length of time and temperature. The results showed that optimum conditions were obtained by single factor experiments, i.e., 240 meshes, 70 °C, 24 h, 12 g sodium silicate (SS), 3 g sodium hydroxide (SH), 1–2 ml Sika (water-soluble plasticizer). The maximum strength properties achieved was 19 MPa. Thus, these materials were equivalent to 20 M grade of cement and can reduce the issues associated with Portland cement (PC).

***Keywords*:** Thermal power plant waste, pond ash, geopolymer, mechanical properties, morphology, thermal stability

**Corresponding author*: muktikanta2@gmail.com

Muktikanta Panigrahi, Ratan Indu Ganguly and Radha Raman Dash (eds.) Development of Geopolymer from Pond Ash-Thermal Power Plant Waste: Novel Constructional Materials for Civil Engineers, (91–122) © 2023 Scrivener Publishing LLC

3.1 Introduction

Coal-based thermal power stations significantly contribute to pollution. It is claimed that 750 million tons of fly ash are generated annually by thermal power stations [1]. Different waste materials are generated due to coal combustion. The generated pollutants are fly ash, bottom ash, pond ash, slags, and flue gas. Fly ash is extracted from flue gases by adopting an electrostatic precipitator or cyclone separator. This also becomes a source for regeneration of valuable metals and can be used as a raw material for the production of ceramics, zeolites, adsorbents, and geopolymers [2–8].

Geopolymers are inorganic polymers that are generated through alkali-activated aluminosilicate. They have adequate strength at ambient temperatures; and since they are environmentally friendly, they can be adopted as suitable structural materials [9, 10]. Bottom ash is collected from the bottom of boilers. It differs from fly ash in terms of particle size range [11, 12]. Although geopolymers are prepared from fly ash, due to its similar composition, it is thought that bottom ash can be used to replace fly ash for preparation of geopolymer [13–15]. Depending on the composition of the coal source, both fly ash and bottom ash may contain heavy metals. In their dry state, they can cause health hazard. Therefore, bottom ash is stored under water. The resulting pond ash may be bottom ash mixed with some amount of fly ash settled under the water in the pond. This pond ash is mixed with soil on land. Since large amounts of ash are disposed of ponds, it has become a threatening proposition for human life. Thus, it is necessary to find a method for utilization of pond ash other than its storage in ponds.

Bottom ash can be collected either in dry state or wet state. Even if it is collected from the pond in a wet state, it contains a significant proportion of fly ash after remaining in the wet state for long periods of time.

Properties of pond ash depend on composition of coal used, burning conditions, time spent in the slurry pond and particle separation during wet storage [16–20]. For this reason, a few studies were conducted on the application of pond ash, including production of geopolymers [21–23]. Reactivity of pond ash depends on size fraction. A fine fraction shows pozzolanic activity and can be used in the production of cement and concrete [24–27]. A coarser fraction of pond ash shows weak pozzolanic activity. This has led to beneficiation using mineral processing techniques to separate the size fractions for possible use as lightweight aggregate [27].

Lee *et al.* used pond ash to produce geopolymer pastes having reasonable strength. They collected pond ashes from a South Korean pond and used it for their work after removal of carbon and drying without any size fractionation [9].

All the above work has not considered the change in the physical properties of pond ashes after storage under wet conditions for a long time. After prolonged storage under water, soluble components of ashes are dissolved, making the material more porous and less dense. Upon treatment of this material with alkali for production of geopolymers, pond ash absorbed large amounts of liquid phase, increasing the liquid to solid ratio of the mixture. This weakens the mechanical property of geopolymers.

The BET (Brunauer, Emmett and Teller) surface areas of pond ashes are up to 25 times higher, and therefore pore volumes are 10 times greater than those of fly ashes from the same power plant [28–31]. Therefore, it is suggested that ashes submerged in ponds have undergone chemical and/or physical changes by leaching out of soluble phases and hydroxylation. Such chemical and physical changes will also depend on how long ashes have been stored under water. Thus, the physical and chemical changes of the pond ash are expected to depend on the duration of storage under water.

Reactivity of pond ash to alkali is influenced by particle size. Grinding can help successfully reduce further particle sizes by grinding operation. Finer particle size enables increased reactivity, which is helpful in producing geopolymer with activated alkali materials [32–34]. Mechanical activation of the pond ash may pave an effective way to utilize pond ashes kept in ash ponds for various duration. However, a drawback to the use of Mongolian pond ash as a viable raw material for the production of geopolymers for construction purposes is the presence of significant amounts of the heavy metals As, Pb and Cr. But it is also known that hazardous elements such as lead can be immobilized in fly ash-based geopolymers by their incorporation into the geopolymer matrix [35–38].

Saxena *et al.* [21] developed innovative strategies to create green concrete with improved strength properties and durability. They prepared geopolymer cement by activating pond ash with alkaline solution (14 M NaOH and sodium silicate solutions). Natural ennore sand was used to prepare geopolymer mortar; and alccofine powder was added during the geopolymerization process, which helped to increase compressive strength of the mortar. Silica fume was also used to prepare geopolymer mortar. Curing at different temperatures was done either in an electric oven or microwave oven. It was observed that microwave-oven curing enabled better compressive strength with shorter curing time. Powder X-ray diffraction, heat evolution profile, TG studies, compression strength, and SEM were used to characterize the geopolymer materials. A durability test in sulfuric acid is considered essential. Durability results showed better stability with microwave-cured material in comparison to other methods of curing.

Kim *et al.* [39] used pond ash to prepare geopolymers. They did so either without adopting any purification process or only with a minimal purification process. Synthetic basalt was used as foaming agent in the preparation of geopolymer. They observed the highest compressive strength (26 MPa) of geopolymers after 7-day curing at ambient temperature. The compressive strength (80 MPa) was thus enhanced considerably by the purification and dry sieving of pond ash. Saxena *et al.* [40] used pond ash containing higher amounts of heavy metals to prepare geopolymer cement. They utilized pond ash fruitfully to prepare geopolymer cement by treating it with alkaline solution followed by thermal curing. Experimental studies were done to assess the activating influences of alkali combinations, i.e., alkali hydroxide and silicate ($NaOH/Na_2SiO_3$, $NaOH/Li_2SiO_3$, KOH/Na_2SiO_3 and KOH/Li_2SiO_3) on the mechanical properties of mortars.

Lee *et al.* [9] characterized pores in geopolymer with the help of X-ray tomography, mercury porosimetry, and gas adsorption method. They observed irregular geometry of pores and approximate equivalent perimeter diameter ranging between 20–60 nm equivalents. Within the volume of 0.00748 μm^3, porosity was determined to be 7.15%. Use of electron tomography was considered to be an important method for measuring the porosity and pore connectivity in geopolymers. Knowledge of porosity helps to correlate structure with properties and predict the durability and properties.

Khodr *et al.* [41] conducted comprehensive studies on geopolymer and geopolymer mortar using brown coal fly ash obtained from two separate locations of an Australian power plant. They obtained good compressive strength with their prepared material.

The above reviews only focused on the recent work that was done in order to present new information on geopolymer production using pond ash. In the following work, the effects of pond ash size, curing time, and curing temperature on the geopolymerization process are studied. The materials are characterized with different characterization techniques and morphological transformations of raw materials to geopolymer are discussed. Different characterization techniques, such as SEM, XRD, TGA, DSC, etc., are used.

3.2 Experimental Details

3.2.1 Materials

3.2.1.1 Pond Ash

Pond ash is a waste product produced by the burning of coal in boilers (Figure 3.1). It is mainly obtained from wet disposal of fly ash, which when

Figure 3.1 Macrograph of pond ash (PA).

mixed with bottom ash is disposed of in large ponds or dykes as slurry. Pond ash is being generated at an alarming rate, which is posing a threat to our environment. Problems are currently being experienced in the management of pond ash disposal, and these have become the thrust area for engineering research.

The chemical composition of pond ash is cited in Table 3.1. It mainly contains silica (SiO_2) and alumina (Al_2O_3), which are expected to take part in the polymerization process. The presence of alkali compounds, such as CaO and MgO, help the process of polymerization.

Table 3.1 Chemical composition of pond ash (PA) [42].

Raw Materials	SiO_2	Al_2O_3	CaO	MgO	Fe_2O_3	TiO_2	Cr_2O_3	MnO	P_2O_5	C	LOI
Pond Ash	62.8	28.3	0.7	0.58	3.85	1.84	0.04	0.03	0.32	1.15	0.5

Table 3.2 Physical properties of pond ash (PA).

SL	Physical properties of PA properties	Pond ash
1	Lime Reactivity of Pond Ash	0.66
2	Specific Gravity	2.16 gm/cc
3	Bulk Density in Loose State	824 kg/m^3
4	Bulk Density in Compacted State	990 kg/m^3
5	Atterberg's Limits Liquid Limits Percentage	47.3
6	IS Classification	SP-SM

3.2.1.2 Physical Properties of Pond Ash

The physical properties of PA [42, 43] are shown in Table 3.2, which indicates the character of the ash used as a major constituent in GP material.

3.2.1.3 Chemicals

Sodium Silicate

Sodium silicate (Figure 3.2) is the common name for compounds with the formula $(Na_2SiO_2)_nO$. A well-known member of this series is sodium metasilicate, Na_2SiO_3. Also known as water glass or liquid glass, these materials are available in aqueous solution and in solid form. The pure compositions are colorless or white, but commercial samples are often greenish or blue owing to the presence of iron-containing impurities [42].

They are used in cements, passive fire protection, textile and lumber processing, refractories, and automobiles. Sodium carbonate and silicon dioxide react when molten to form sodium silicate and carbon dioxide. The open chemical structure of sodium silicate is shown in Figure 3.2.

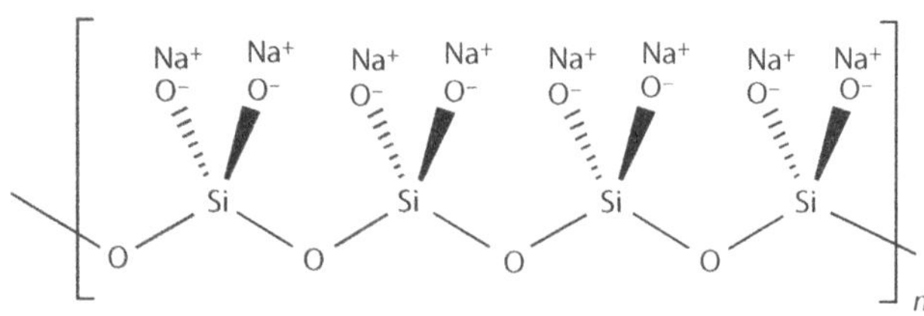

Figure 3.2 Chemical structure of sodium silicate.

Sodium Hydroxide

Sodium hydroxide is also known as lye or caustic soda, which is an inorganic compound with the formula NaOH [44]. It is a white solid ionic compound consisting of sodium cations Na+ and hydroxide anions OH−. Sodium hydroxide is a highly caustic base and alkali that decomposes proteins at ordinary ambient temperatures and may cause severe chemical burns. It is highly soluble in water, and readily absorbs moisture and dioxide from the air. It forms a series of hydrates NaOH*n*.H2O. https://en.wikipedia.org/wiki/Sodium_hydroxide - cite_note-siem-10. The monohydrate $NaOH{\cdot}H_2O$ crystallizes from water solutions between 12.3 and 61.8 °C. The commercially available "sodium hydroxide" is often this monohydrate, and published data may refer to it instead of the anhydrous compound. As one of the simplest hydroxides, it is frequently utilized alongside neutral water and acidic hydrochloric acid to demonstrate the pH scale to chemistry students. Sodium hydroxide is used by many industries to manufacture pulp and paper, textiles, drinking water, soaps and detergents, and as a drain cleaner. Figure 3.3 shows a photograph sodium hydroxide pellets.

Sika® (Water-Soluble Plasticizer)

Sika is a highly concentrated air-entraining liquid plasticizing admixture which is added to strengthen products. It forms stable air bubbles in mortar mixes. This improves the workability of mortar, giving a buttery consistency, and helps to prevent shrinkage, cracking, crazing, and mild frost damage during the curing process as well as providing long-term resistance to freeze/thaw cycles. It also replaces lime in the mix, and hence

Figure 3.3 Sodium hydroxide pellets.

Figure 3.4 Sika (water-soluble plasticizer).

reduces the chance of efflorescence (salt formation) and smooths out variations between separate batches of sand and cements. A photograph of Sika [42, 44] is shown in Figure 3.4.

3.2.2 Preparation of Geopolymer from Pond Ash

The synthesis of geopolymer is carried by the solid-state reaction route via chemical process. Geopolymer and geopolymeric products (mortar and concrete) are prepared using different components such as pond ash, high-carbon ferrochrome slag, jute fiber, sand, and aggregates with alkali-activated solution. The process of preparing different geopolymeric materials are discussed step-by-step below. The process is carried out in four steps, i.e., grinding, mixing, ramming, and curing in a hot air oven [42, 44].

Step 1: Raw Material Preparation

Pond ash-based geopolymer is prepared by taking a requisite amount of as-received pond ash, sodium silicate, sodium hydroxide, and water-soluble plasticizers. The as received pond ash from NALCO Damanjodi is ground by ball mill followed by sieving to different mesh sizes (e.g., 240 meshes, 200 meshes and 150 meshes).

Step 2: Geopolymer Preparation with Raw Materials

The 240 mesh size of pond ash is mixed with a requisite amount of NaOH, Na_2SiO_3, water and water-soluble plasticizers (Sika). A gel-like mass is formed. Three samples of same compositions are chosen (Table 3.3) in order to take average response of the mixtures.

Step 3: Molding, Casting and Compaction

An appropriate amount of gel-like paste is poured into a REMI iron mold of 50 mm diameter and 70 mm height. The samples are rammed continuously

Table 3.3 PA-based geopolymer mixtures (three samples of same compositions).

Sample code	PA	Sod. Silicate (SS)	Alkali (8 M, NaOH) (SH)	Water soluble Plasticizer (Sika)
S1	85%	12%	3%	1-2ml
S2	85%	12%	3%	1-2ml
S3	85%	12%	3%	1-2ml

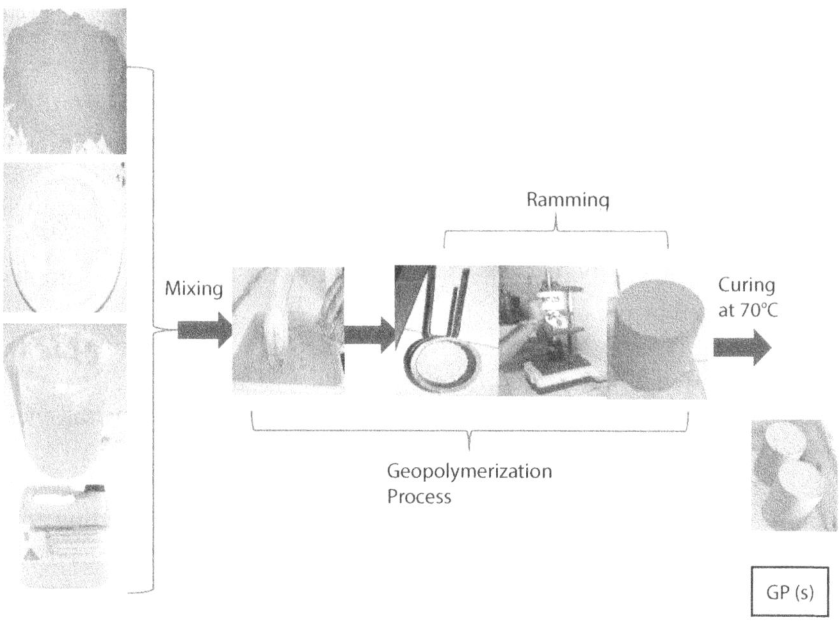

Figure 3.5 Schematic diagram of geopolymer preparation process of pond ash.

twenty (20) times for dense compaction. They are then left for 3 minutes before demolding. These samples are designated as "green" samples.

Step 4: Curing

The cast green samples are cured at 70 °C for 24 h in an oven. The cured samples are preserved in plastic zipper bags until they are tested.

Geopolymer samples with different compositions and mesh sizes are prepared by this process using different input variables such as temperature, time, and mesh size for optimization purpose. A schematic diagram of the entire geopolymer preparation process of pond ash is shown in Figure 3.5.

3.2.3 Test Methods

Fourier-transform infrared spectroscopy (FTIR) is a non-destructive testing (NDT) technique. The presence of functional groups, formation of chemical linkage between used materials and interaction between nanomaterials and polymers, metal oxygen bond and removal of organic and other phases have been investigated using a Thermo Nicolet NEXUS 870 FTIR spectrometer (Thermo Fisher, USA). The instrument parameters were kept constant (50 scan at 4 cm^{-1} resolution, transmittance/absorbance mode). In this spectrometer [45, 46], the IR radiations from an IR source are passed through the sample and the amount of energy adsorbed/transmitted was recorded by suitable detector and guided through an interferometer where a Fourier transform is performed on the output signal.

For this technique, powder, liquid and solid film samples are used. Solid film samples are analyzed in ATR mode. For this measurement, the powdered samples are prepared by making pellets. First, the dry KBr samples are ground using a mortar pestle followed by the addition of a small amount of prepared samples. Then the sample mixture is ground further and put into a die and placed inside the hydraulic press, where the samples are compacted by pressure (5 kgf).

The pellet (13 mm diameter, 0.3 mm thick) so prepared is used for IR characterization. Liquid samples are directly put on KBr pellet/quartz glass plate and used for this characterization. Before running the samples, a background spectrum is collected. Then pellet samples are put in a sample holder. The pellets are exposed to IR radiation in the spectrometer and data are collected. This technique is done to characterize the bonding type of the molecules and for each type of bonding it produced characteristic absorption bands.

X-ray diffraction (XRD) is a powerful non-destructive technique to determine crystalline/amorphous structure [47] and intercalated/

exfoliated/delaminated nanostructure [47] of clay-based composites. It also estimates various structural parameters such as crystallite size (D), interlayered-spacing (d), etc. X-rays are generated from Cu target with a characteristic wavelength (λ) which can be obtained from the expression [48, 49], as shown in Equation (3.1),

$$E = h\upsilon = h\frac{c}{\lambda} \tag{3.1}$$

where h is the Planck's constant (6.62×10^{-34} joule), c is the velocity of light (3×10^{-8} m/s) and E is the energy of the radiation. The wavelength of X-ray is comparable to the size of atoms.

The basic principle of X-ray diffraction is based on constructive interference of X-rays (monochromatic) and prepared samples. The interaction of the incident X-rays with the samples produces constructive interference and the diffracted rays are generated, which satisfies Bragg's law ($n\lambda = 2d\sin\theta$). By scanning the sample through a range of 2θ angles, all possible diffraction directions of the lattice should be attained due to the random orientation of the powdered materials. Conversion of the diffraction peaks to d-spacing allows the identification of the materials because each material has a set of unique d-spacing. Typically, this is achieved by comparison of d-spacing with standard reference patterns.

In the present work, most of the X-ray diffraction patterns were recorded using CuKα radiation (wavelength, λ = 0.154 nm). During the operation, XRD was operated at 40 kV and 20 mA. The powder samples were placed on a quartz sample holder at room temperature and were scanned at diffraction angle 2θ from 5° to 45°. The sample is kept in a Perspex holder.

Surface morphologies of prepared materials are analyzed by electron microscopy techniques such as SEM/FESEM. An SEM/FESEM microscope uses electrons in place of light to produce an image [50]. It is a surface phenomenon. In this measurement, the electron beam produced from an electron gun is focused on a small portion of the sample that is kept in a vacuum. A detector collects the output signals during the interaction of electrons with the sample, which are sent to a computer. This forms the final image. Two types of electron guns are used; one is thermionic and the other is field emission.

The electrons emitted from the electron gun are accelerated by applying a high electric potential. The scanning of the electron beam over the sample surface is controlled by deflecting the electron beam using a scanning coil. In vacuum condition, both FESEM and SEM are used to produce images. Therefore, a special preparation technique is needed for the sample to avoid moisture absorption. All non-conducting materials need a thin layer of conducting coating. This is done by a "sputter coater" with an operating voltage of 4 kV. Such a coater uses an electric field and argon gas. The sample is placed in a small vacuum chamber. The argon gas is ionized in the applied electric field to form argon ion (Ar^+). The argon ions knock gold atoms from the surface of the gold foil and get deposited on sample.

The coupling of DSC and TGA signal is advantageous because it measures physical transformation as evaporation, sublimation, drying, the physical property as heat flow, as well as the chemical properties required for the kinetics. The TGA curve of materials shows the mass loss steps relating to the loss of volatile components, materials decomposition, combustion and final residues. Based on this information, it is even possible to conclude the individual components of the sample, to some extent. Furthermore, it offers the first derivative of the TGA curve, the DTG curve, the relation to time and proceeds proportional to the rate of decomposition. The combination of both methods offers. Furthermore, artefacts have to be taken into account and those effects on the measurement curve which are not caused by the sample. Nominal valued artefacts are the occurrence effect, which is caused by the gas density decreasing during heating; a weight increase is the consequence. This effect can be counteracted by a blank curve [51]. A blank curve, or a base line, is a measurement curve which has been created under the same conditions, but without a sample, so only an empty crucible [51]. Fluctuations in the flushing gas stream are another artefact since they can be seen on the measurement curve and should therefore not be changed during the measurement [51, 52].

The TGA sample carrier cannot measure a DSC signal. Before starting with the analysis, the first step is to calibrate the sample carrier for sensitivity and temperature. This calibration occurred with aluminium. All of the samples were heated 50 °C above the melting point and then cooled 50 °C below it; this procedure was repeated five times. The observed onset of the melting point temperature and enthalpy (area underneath the curve) were afterwards compared with the existing values of the mentioned metals.

The selection of the crucibles is crucial since it can influence the measurement in various ways. First of all, the size of the crucible determines the amount of the sample. It can also reduce the amount of resolving power through the heat capacity of the crucible. The material of the crucible must not react with the sample; otherwise, the measurement will be disturbed and thus unusable [51, 52]. The chosen crucible has a volume of 20 micrometers, made of aluminium oxide (Al_2O_3), and the sample weight is therefore 10 milligrams (weight with a tolerance of +/- 0.1%).

The pyrolysis process occurs in an oxygen-free atmosphere; therefore, it is necessary to use inert gas such as argon. All tests had the same temperature program, starting with 34 °C up to 800 °C, and run through a start-standby (heating rate for 10°/min) to observe the drying and to better compare the different runs.

Before starting with the first sample, it was necessary that the empty crucible run through a whole test program in order to afterwards correct the curves from possible deviation triggered from the crucible. This crucible correction creates the blank curve and was always repeated for all heating rates before the measurements with the samples. Between all pyrolysis tests the furnace needed an oxidative process (800 °C, Argon gas) to remove all possible remaining sample in the furnace.

Differential scanning calorimetry (DSC) is a technique for measuring the energy necessary to establish a nearly zero temperature difference between a substance and an inert reference material, as the two specimens are subjected to identical temperature regimes in an environment heated or cooled at a controlled rate. The technique provides qualitative and quantitative information about physical and chemical changes that involve endothermic or exothermic processes or changes in heat capacity using minimal amounts of sample. It has many advantages, including fast analysis time (typically thirty minutes), easy sample preparation, applicability to both liquids and solids, a wide range of temperature applicability and excellent quantitative capability [52].

There are two types of DSC systems commonly used. In power-compensated DSC, the temperatures of the sample and reference are controlled independently using separate, identical furnaces. The temperatures of the sample and reference are made identical by varying the power input to the two furnaces; the energy required to do this is a measure of the enthalpy or heat capacity changes in the sample relative to the reference.

The DSC system has been used in the evaluation of small transitions, such as multiple phase transitions in liquid crystals and those due to side chains in polymers, which cannot be resolved by most other techniques. It allows accurate determination of temperatures associated with thermal events. Temperature can be calibrated with respect to one or more standards, which allows highly accurate, precise and reproducible values. The technique reveals the thermal history imparted to thermoplastics as a result of different processing conditions. The information generated can be used to vary heating rates to deliver the required degree of crystallinity.

A differential scanning calorimeter (Diamond DSC, Perkin-Elmer, USA) was used for determination of crystallization. Samples (5–10 mg) were placed in sealed aluminium pans and scanned under a constant nitrogen purge (20 mL/min). Subsequently, the samples were heated from 30 °C to 200 °C at a rate of 20 °C/min, held at 200 °C for 2 min, cooled to 30 °C at the same rate and held for 2 min to stabilize. Finally, a second scan was carried out from room temp to 200 °C. The resulting crystallization temperature (T_c), melting temperature (T_m) and glass transition temperature (T_g) were noted from the second scan.

In order for geopolymer to be a novel cementitious material, its mechanical properties, such as compressive strength, tensile strength, flexural strength and bond strength, are very important additions to concrete. Its better mechanical properties over PC have been perceived as advantages in previous studies [63, 64]. It has been shown that compressive strength, flexural strength, and tensile strength of geopolymers increase as NaOH solution concentration increases [65].

Compressive strength was tested on the cylindrical samples (100 mm diameter and 200 mm depth) of concrete and cubic samples (50×50×50 mm^3) of mortar. The testing procedure for compressive strength was based on the AS 1012.9 [66]. All compressive strength tests were conducted using the Controls MCC (Multifunctional Computerized Control Console) machine of 3000 kN capacity. It was electronically operated and the data were acquired from its software. Top surface of cylinders was capped using rubber capping plate to distribute load evenly on the top loading surface of the sample. The loading rate was kept fixed as 0.333 MPa/sec (approximately 20 ± 2 MPa/min) for compressive strength test for both mortar and concrete samples. Load was applied gradually on the specimen until failure. The machine recorded the maximum force exerted and calculated the compressive strength using the given values of cylinder dimensions that was provided beforehand.

3.2.4 Results and Discussion

Different treatment combinations of variables are chosen systematically to optimize variables for obtaining the highest compressive strength (CS) of PA-based geopolymer. The following tables are provided to show the effect of variables on strength properties.

Table 3.4 shows the variations of sodium silicate (SS) and alkali solutions (SH=Sodium hydroxide) from 15–12% and 3–1%, respectively, while other variables, i.e., pond ash (85%) and Sika (1–2 mL), are kept constant. The strength of geopolymer prepared with three compositions is shown in Table 3.4 for curing temperature (70 °C) and two curing times (1 day and 7 days). The maximum strength is obtained by curing for 7 days for

Table 3.4 Variation of SS and SH on mechanical properties (curing time 24 h, curing temperature 70 °C, mesh 240).

Sample code	Pond ash	Sod. silicate (SS)	Alkali (8 M, NaOH) (SH)	Water-soluble plasticizer (Sika)	Compressive strength (MPa) (1 day curing)	Compressive strength (MPa) (7 days curing)
*S1	85%	12%	3%	1–2 ml	20	21.0
*S2	85%	13%	2%	1–2 ml	19	20.4
*S3	85%	14%	1%	1–2 ml	18	19.7

*__Note:__ Each value is average from 10 samples.

treatment combination of PA (85%), 12% SS, 3% SH, and Sika (1–2 mL). Maximum strength is found to be 21.0 MPa.

Table 3.5 shows the variations of mesh sizes from 150 mesh, 200 mesh, and 240 mesh, respectively, while other variables, i.e., pond ash (85%), sodium silicate (12%), sodium hydroxide (3%), and Sika (1–2 mL) are kept constant. Strength of geopolymer prepared with compositions is shown in Table 3.5 for curing temperature (70 °C) and curing time (24 h). The maximum strength is obtained by curing for 24 h for treatment combination of PA (85%), 12% SS, 3% SH, and Sika (1–2 mL). Maximum strength is found to be 19.0 MPa.

Table 3.6 shows variations in curing temperatures (50 °C, 60 °C, 70 °C, and 80 °C); while other variables, i.e., pond ash (85%), sodium silicate (12%),

Table 3.5 Variation of mesh size on mechanical properties (curing time 24 h, curing temperature 70 °C, 12% SS and 3% SH).

Sample code	Mesh size (micron size)	Pond ash	Sod. silicate	Alkali (8 M, NaOH)	Water-soluble plasticizer (Sika)	Compressive strength (MPa)
*S1	150	85%	12%	3%	1–2 ml	15
*S2	200	85%	12%	3%	1–2 ml	16.5
*S3	240	85%	12%	3%	1–2 ml	19

*__Note:__ Each value is average from 10 samples.

Table 3.6 Variation of curing temperatures on mechanical properties (curing time 24 h, 240 mesh size, 12% SS and 3% SH).

Sample code	Curing temp. (°C)	Pond ash	Sod. silicate	Alkali (8 M, NaOH)	Water soluble plasticizer (Sika)	Compressive strength (MPa)
*S1	50	85%	12%	3%	1–2 ml	15
*S2	60	85%	12%	3%	1–2 ml	16.5
*S3	70	85%	12%	3%	1–2 ml	19
*S4	80	85%	12%	3%	1–2 ml	15

***Note:** Each value is average from 10 samples.

sodium hydroxide (3%), Sika (1–2 mL) and curing time (24 h) are kept constant. Strength of geopolymer prepared with compositions is shown in Table 3.6 for two curing times (1 day and 7 days). The maximum strength is obtained by curing for 1 day for treatment combination of PA (85%), 12% SS, 3% SH, and Sika (1–2 mL). Maximum strength is found to be 19.0 MPa.

Table 3.7 shows variations in length of curing times from 4 h, 8 h, 16 h, 20 h, 24 h, and 168 h, respectively; while other variables, i.e., pond ash (85%), Sika (1–2 ml), sodium silicate (12%), sodium hydroxide (3%), are kept constant. Strength of geopolymer prepared with compositions having different lengths of curing time is shown in Table 3.7 for curing temperature (70 °C). The maximum strength is obtained by curing for 168 h for treatment combination of PA (85%), 12% SS, 3% SH, and Sika (1–2 ml). Maximum strength is found to be 20.3 MPa.

It is concluded from the results of experiments shown in Table 3.4, Table 3.5, Table 3.6 and Table 3.7, respectively, that the best compressive strength property is obtained for a treatment combination of 12% Na_2SiO_3, 3% NaOH (8 M), 70 °C curing temperature and 24 h curing time. To achieve the highest strength, 8 time ramming is recommended. Maximum achievable strength is 20–21 MPa.

All the strength properties are evaluated at room temperature for all the experiments. However, it is considered useful to find the strength behavior of GP material at higher temperature for structural integrity with the materials. For testing GP at higher temperature, a special fixture is designed

Table 3.7 Variation of curing times on mechanical properties (curing temperature (70 °C), 240 mesh size, 12% SS and 3% SH).

Sample code	Curing time (hour)	Pond ash	Sod. silicate	Alkali (8 M, NaOH)	Water-soluble plasticizer (Sika)	Compressive strength (MPa)
*S1	4	85%	12%	3%	1–2 ml	10.6
*S2	8	85%	12%	3%	1–2 ml	13
*S3	16	85%	12%	3%	1–2 ml	15
*S4	20	85%	12%	3%	1–2 ml	18
*S5	24	85%	12%	3%	1–2 ml	19
*S6	168	85%	12%	3%	1–2 ml	20.3

***Note:** Each value is average from 10 samples.

to test GP samples at 90 °C. The designed grip made of stainless steel is attached to the moveable jaw of the universal tensile testing machine.

The test samples are placed in a holder fixed with the attachment where temperature (Figure 3.6) is controlled with an accuracy of ±1 °C. Testing is conducted at a slow strain rate for GP samples which are prepared using optimum process parameters (70 °C, 24 h, and 240 meshes). The results cited in Table 3.8 show:

1) The strength value is comparable to the strength obtained for the materials at ambient temperature. Thus, the structural integrity is maintained even at a higher temperature.
2) Increasing curing time from 1 day to 7 days does not affect the strength of the materials.

The SEM images for as-received pond ash and pond ash-based geopolymers are shown in Figure 3.7. It is observed that for as-received samples there are irregular shaped particles that are relatively loose, indicating high porosity; whereas prepared pond ash-based geopolymer Figures 3.7b, 3.7c, 3.7d, 3.7e and 3.7f is compacted mass. Figure 3.7 shows SEM images of GP samples cured for different lengths of time. It can be observed that there is progressive polymerization in the materials during curing for different lengths of time at 70 °C. Agglomeration of loose particles (Figure 3.7a)

Figure 3.6 MOR set-up (fabricated) for elevated temperature studies.

Table 3.8 Mix proportions of pond ash-based geopolymer cured for 1 day and 7 days at 70 °C.

Pond ash	Sod. silicate	Alkali (8 M, NaOH)	Water-soluble plasticizer (Sika)	Compressive Strength (MPa) [24 h]	Compressive strength (MPa) [24 h] at 90 °C	Compressive strength (MPa) [7 days curing] at 90 °C
85%	12%	3%	1–2 ml	19.0	20.3	21.0

can be observed in Figures 3.7b, 3.7c, 3.7d, 3.7e and 3.7f. Figures 3.7a and 3.7c are micrographs of GP cured at 50 °C and 60 °C respectively. Agglomerations due to curing at those temperatures are well documented in the two micrographs (Figures 3.7a and 3.7c). Elements present in GP are marked in Corresponding EDX analysis (Figures 3.7b and 3.7d). In those areas, elements of Si, AL, Fe, Mg, Ca, Na, O, etc., have been identified.

Comparison of micrographs (Figures 3.8a and 3.8c), obtained from two GP samples cured at two different temperatures of 50 °C and 60 °C for the same length of time indicate that curing at 60 °C has enabled greater agglomeration of powder materials due to increased polymerization at a higher temperature. Similar micrographs of GP at two other temperatures (i.e., 70 °C and 80 °C, respectively) for the same length of curing time are

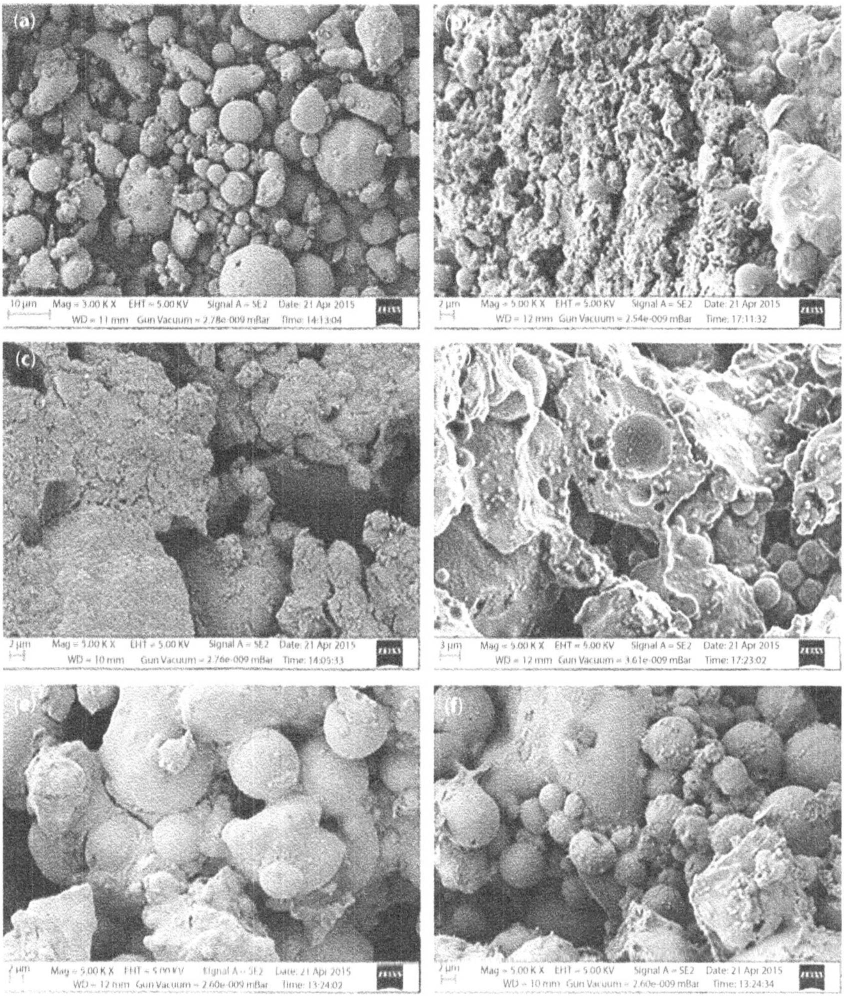

Figure 3.7 SEM images of (a) As-received Pond Ash; (b) PA-based GP cured at 70 °C for 4 h; (c) PA-based GP cured at 70 °C for 8 h; (d) PA-based GP cured at 70 °C for 16 h, (e) PA-based GP cured at 70 °C for 20 h, and (f) PA-based GP cured at 70 °C for 24 h.

shown in Figures 3.8e and 3.8g. The same characteristics of agglomerated GP are observed. The striking feature of the micrograph shown in Figure 3.8g (region-marked) is the presence of needle-shaped phases (marked), wide-spread on the work surface of the particles. The needles have high aspect ratios. When the microstructures are examined by EDX analysis (Figure 3.8b), it confirms the presence of elements such as aluminium,

Figure 3.8 SEM images of (a) PA-based GP cured at 50 °C, (b) EDS of PA-based GP cured at 50 °C, (c) PA-based GP cured at 60 °C, and (d) EDS of PA-based GP cured at 60 °C. *(Continued)*

silicon, and oxygen which super of formation of geopolymer with Al, O, Si, etc.

The presence of Si, Al, Fe, Ca, Mg, Na, and O is revealed by EDX analysis. The EDX data of as-received pond ash and as-prepared PA-based GP at different temperatures for same curing time is shown in Figures 3.8b, 3.8d, 3.8f, and 3.8h. The Si/Al ratio has been observed to be 2.11 for as-received

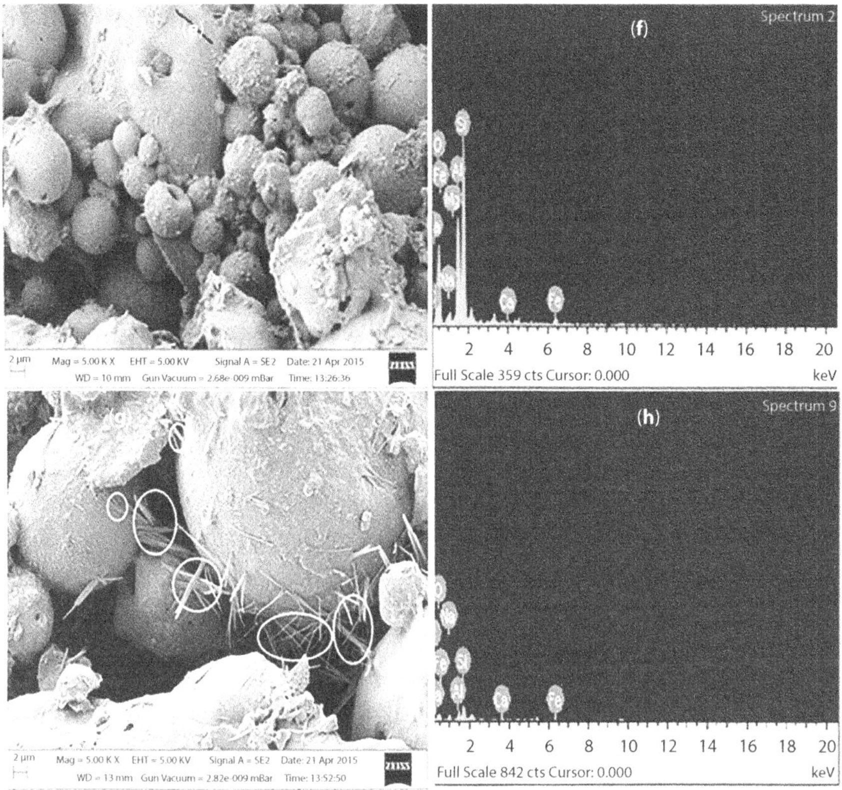

Figure 3.8 (Continued) SEM images of (e) PA-based GP cured at 70 °C, (f) EDS of PA-based GP cured at 70 °C, (g) PA-based GP curried at 80 °C, and (h) EDS of PA-based GP cured at 80 °C.

PA, and after curing at 70 °C for 24 h, the ratio changes to 3.38. From these results, it can be said that increase in Si/Al ratio has aided in increasing the compressive strength of the cured geopolymers. This may happen due to the formation of polymer chain. The gel phase surrounding the particles (Figures 3.8a, 3.8c, and 3.8e) indicates that the mechanical strength of the geopolymeric binder is higher than that at lower curing time and temperature.

Geopolymer pastes are characterized by DSC. Using this technique, it is possible to observe exothermic and endothermic events, as well as glass transition temperatures (*Tg*). The range of investigation is between 30–200 °C. Results obtained from DSC thermograms (exothermal peaks) of the as-received PA and PA-based geopolymer at different curing times are shown in Figure 3.9 The DSC thermogram of as-received PA is straight, indicating no polymerization reaction. The thermograms of as-prepared geopolymer show a single peak, indicating the degree of geopolymerization [53, 54].

Geopolymerization occurred by condensation polymerization, which produced some water molecules. Ocular water molecules started oozing out from 70 °C onwards till ~110 °C and are marked in Figure 3.9 as endotherms. Peaks are also observed for all GP samples cured for different lengths of time. Within the temperature range of 30–200 °C, the peak position and peak area of the DSC thermogram of as-prepared geopolymer are different. This indicates a difference in the degree of geopolymerization of prepared geopolymers at different process variables.

Samples of cured GP are characterized using TG-DTG isothermal analysis (Figure 3.10). The significant observation made is loss of water beginning around 50 °C till 400 °C temperature (humps in the thermogram), indicating loss of water since polymerization is completed during curing [55–59].

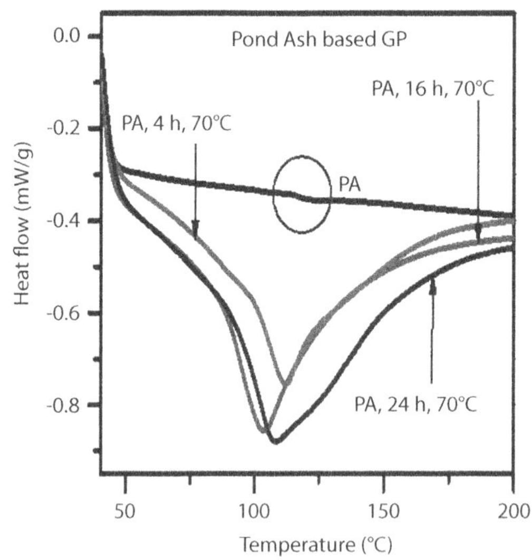

Figure 3.9 DSC isotherm of pond ash-based GP at different lengths of curing length.

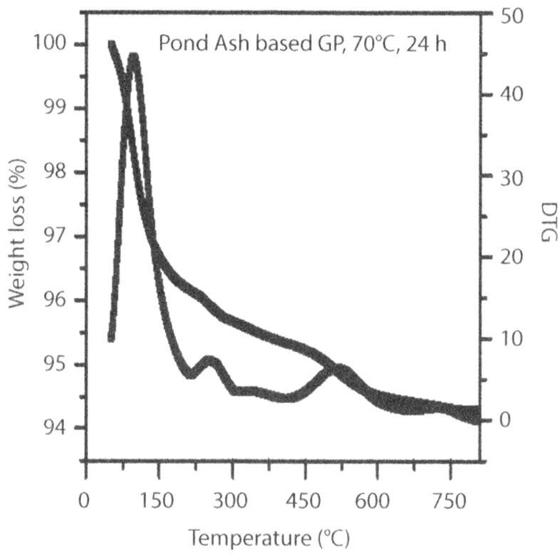

Figure 3.10 TG/DTA analysis of PA-based GP (at 70 °C in 24 h).

The FTIR spectra for pond ash and its geopolymer pastes are shown in Figure 3.11a and 3.11b; and peak assignments with corresponding wave number are presented in Table 3.9. The main feature of the FTIR spectra was the central band at around 1093 cm^{-1}, which is attributed to the Si–O–Si or Al–O–Si asymmetric stretching mode [60]. The main spectral band originally appearing in pond ash at about 1078 cm^{-1} shifted to lower frequencies after geopolymerization took place. The larger the shift, the

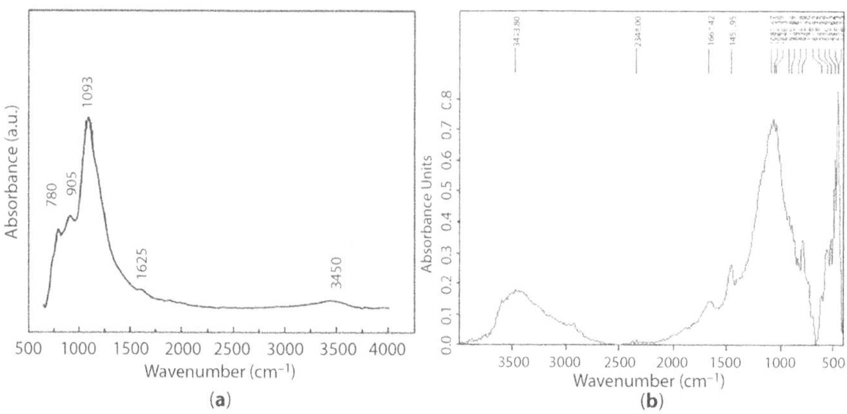

Figure 3.11 (a) FTIR of sieved as-received PA; (b) FTIR of PA GP (cured at 70 °C in 24 h).

Table 3.9 Wave number and peak assignments of as-received PA and PAGP.

S. no.	Wave number		Peak assignments
	As-received PA	**PA GP**	
1	3450	3473	O-H stretching band
2	1625	1662	O-H bending band
3	---	2342	
4	---	1452	Asymmetric carbonate stretching
5	1093	1087	Si–O–Si or Al–O–Si asymmetric stretching
6	905	919	Mullite band
7	780	835	Quartz

higher the degree of penetration of Al from the glassy part of the pond ash into the $(SiO_4)^{4-}$ net. This indicates that the geopolymerization process is influenced by both parameters. Significant broad bands were observed at approximately 3450 and 1640 cm^{-1} for the O–H stretching mode and O-H bending mode [61, 62].

The presence of quartz is shown by a characteristic doublet at around 796 and 776 cm^{-1} that is present in all samples. Another spectral band at around 1440 cm^{-1} appeared in all the geopolymer samples, but is absent in the pond ash. This band is characteristic of the asymmetric CO_3 stretching mode, which suggests the presence of sodium carbonate as a result of the reaction between excess sodium and atmospheric carbon dioxide. Excess sodium content can form sodium carbonate by atmospheric carbonation and may disrupt the polymerization process. Particularly in the case of low-Ca alkali-activated materials, when sodium carbonate reacts with CO_2 it tends to form sodium carbonates and bicarbonates, which are more soluble than the $CaCO_3$ formed by reaction of OPC and may therefore act as an alkali sink and/or play a buffering role in the solution [63, 64].

3.3 Conclusions

In the first phase of this work, PA procured from NALCO, Damanjodi, Odisha, was graded into different fractions through screening (i.e., sieving).

An adequate amount of different fractions of graded pond ash was mixed with a requisite amount of moisture and other chemicals as described in this text; and the mixture was cast in a mold. The cast materials were then cured. Parameters such as size, curing time, and temperature were optimized to get the best product strength. The maximum strength achieved was comparable to that of M15-M20 grade conventional cement-based products. The integrity of structure was verified by testing the materials up to 90 °C. A regression equation was developed to quantify the effect of parameters on strength properties. This equation can predict strength of the material within the range of variations of variables. The adequacy of the equation was statistically checked and was capable of predicting the properties with 95% confidence. GP formed with PA showed strength equivalent to the M15 and M20-grade PC used for structural applications, ensuring the integrity of a structure up to a temperature of 70 °C.

The material was characterized with the help of SEM/EDS, XRD, DSC, TGA, and FTIR respectively. The mechanism involved is that silicon and aluminium atoms present in waste reacts with alkali liquid, forming geopolymer which binds other non-reactive materials in the waste. The highest strength level achieved for as-prepared geopolymers was 19 MPa after curing 24 h. The SEM micrographs reveal that there was a gradual transformation from irregular spherical shape to compacted mass, which was due to polymeric transformation with increase in curing time, which also was corroborated with mechanical properties such as compressive strength. Distinct changes were observed in FTIR spectra, i.e., increase in peak height as well as appearance of many other peaks if compared with FTIR spectrum of virgin material. This strength was due to polymeric reactions and formation of chains with the monomeric structure. A few interesting features of the TGA pattern were observed for the two materials, i.e., uncured and cured samples, described in the text. DSC isotherms indicated oozing out of inbuilt water which accumulated during condensation polymerization reaction. The strength level achieved by optimum use of variables was comparable to that of standard M15 grade mortar used for constructional purpose.

Acknowledgments

The authors convey their sincere thanks to the Ministry of Mines, Government of India, for providing the grant to carry out their work [Grant number= F.No.:14/54/20214-Met-IV dated: 29.12.2014]. They also convey their sincere thanks to GIET, Gunupur, Rayagada, Odisha, India, for providing lab facilities to do the research work. The authors also thank the CRF, IIT Kharagpur, for providing testing facilities.

References

1. A. Rastogi and V.K. Paul, (2020), A Critical Review of the Potential for Fly Ash Utilization in Construction-Specific Applications in India, Journal of Environmental Research, Engineering and Management Vol. 76, pp. 65-75 (https://doi.org/10.5755/j01.erem.76.2.25166)
2. International Energy Agency (IEA), Coal-Fired Power, Tracking Clean Energy Progress (https://www.iea.org/tcep/ power/coal/).
3. International Energy Agency (IEA), Beta. International. (https://www.eia.gov/beta/international/).
4. United States Energy Information Administration. Coal explained. Use of coal. (https://www.eia.gov/ energyexplained/coal/use-of-coal.php).
5. United States Energy Information Administration. Frequently asked questions. What is U.S. electricity generation by energy source? (https://www.eia.gov/tools/faqs/faq. php?id=427&t=3).
6. United States Energy Information Administration. Today in Energy. Chinese coal-fired electricity generation expected to flatten as mix shifts to renewables (https://www. eia.gov/todayinenergy/detail.php?id=33092).
7. J. Chen, G. Liu, Y. Kang, B. Wu, R. Sun, and C. Zhou, (2014), Coal Utilization in China: Environmental Impacts and Human Health, Environ Geochem Health, Vol. 36, pp. 735-53.
8. S. Kumar, J. Singh, and S.K. Mohapatra, (2018), Role of Particle Size in Assessment of Physico-Chemical Properties and Trace Elements in Indian Fly Ash, Waste Mang. Res, Vol. 36, pp. 1016-22.
9. S. Lee, H.-T. A. Riessen, William D.A.Rickard, C.-M. Chon, N.-H. Kang, (2014), Three-Dimensional Quantification of Pore Structure in Coal Ash-based Geopolymer using Conventional Electron Tomography, Construction and Building Materials, Vol. 52, pp. 221-226 (https://doi.org/10.1016/j.conbuildmat.2013.10.072).
10. Tianyu Xie, Togay Ozbakkaloglu, (2015), Behavior of Low-Calcium Fly and Bottom Ash-based Geopolymer Concrete Cured at Ambient Temperature, Ceramics International, Vol. 41, pp.5945-5958.
11. P.-C. Chiang and S.-Y. Pan, (2017), Fly Ash, Bottom Ash, and Dust Carbon Dioxide Mineralization and Utilization, Book Chapter, pp. 253-264.
12. D. Kumar1, Neetesh Kumar2, Ashish Gupta3, (2014), Geotechnical Properties of Fly Ash and Bottom Ash Mixtures in Different Proportions, International Journal of Science and Research (IJSR), Vol. 3, pp. 1487-1494.
13. L.J. Jaramillo, N. Fábio, E. Silviany, G. Marianade, S.P. Erick, Z.V. Adriano, and M. Bernardin, (2021), Use of Fly and Bottom Ashes from A Thermoelectrical

Plant in the Synthesis of Geopolymers: Evaluation of Reaction Efficiency, Energy Geoscience, Vol. 2, pp. 167-173.

14. A.B. Abulencia, M.B.D. Villoria, R.G.D. Libre, P.R.J. Quiatchon, I.J.R. Dollente, E.J. Guades, M.A.B. Promentilla, L.E.O. Garciano, and J.M.C. Ongpeng, (2021), Geopolymers as Sustainable Material for Strengthening and Restoring Unreinforced Masonry Structures: A Review, Buildings, Vol. 11, pp.1-27, (https://doi.org/10.3390/buildings11110532).
15. K. Boonserm, V. Sata, K. Pimraksa, and P. Chindaprasirt, (2012), Improved Geopolymerization of Bottom Ash by Incorporating Fly Ash and using Waste Gypsum as Additive, Vol. 34, pp. 0-0. (DOI:10.1016/j.cemconcomp.2012.04.001).
16. A.K. Bera, A. Ghosh, and A. Ghosh, (2017), Compaction Characteristics of Pond Ash, Journal of Materials In Civil Engineering, Vol. 19, pp. 349-357 (DOI: 10.1061/(ASCE)0899-1561 (2007)19:4(349)).
17. K. Arumugam, R. Ilangovan, and D.M. James, (2011), A Study On Characterization And Use of Pond Ash as Fine Aggregate in Concrete, International Journal Of Civil And Structural Engineering Vol. 2, pp. 466-474.
18. J. Mallick and D.P. Palai, (2016), Collapsible Behavior of Pond ash, International Research Journal of Engineering and Technology (IRJET), Vol. 3, pp. 480-485.
19. M. Suthar and P. Aggarwal, (2018), Bearing Ratio and Leachate Analysis of Pond Ash Stabilized with Lime and Lime Sludge, Journal of Rock Mechanics and Geotechnical Engineering, Vol. xxx, pp. 1-9 (https://doi.org/10.1016/j.jrmge.2017.12.008)
20. M. Suthar and P. Aggarwal, (2016), Environmental Impact and Physicochemical Assessment of Pond Ash for its Potential Application as a Fill Material, Int. J. of Geosynth. and Ground Eng., Vol. 2, pp.1-9 (DOI 10.1007/s40891-016-0061-7).
21. S.K. Saxena, M. Kumar, N.B. Singh, (2018), Effect of Alccofine Powder on the Properties of Pond Fly Ash based Geopolymer Mortar under Different Conditions, Environmental Technology & Innovation, Vol. xxx, pp. 1-20 (DOI: https://doi.org/10.1016/j.eti.2017.12.010).
22. J. Temuujina, A. Minjigmaaa, U. Bayarzula, D.S. Kimb, S-Ho Leeb, H.J. Leeb, C.H. Ruescherc, K.J.D. MacKenzied, (2017), Properties of Geopolymer Binders Prepared from Milled Pond Ash, Materiales de Construcción Vol. 67, pp. 1-20 (http://dx.doi.org/10.3989/mc.2017.07716).
23. R.V. Ranganath, L.R. Manjunath, B.H. Nishanth and V. Vidyadhara, (2021), Properties of Geopolymer Composites Using Activated Pond Ash, Vol. xxx, pp. 1-10.
24. A. Verma, A. Kumar, A. Mishra, and A. Verma, (2016), Utilization of Pond Ash as Partial Replacement of Cement Concrete Mix, International Research Journal of Engineering and Technology (IRJET), Vol.3 pp. 511-515.

25. P.P. Bhangale and P.M. Nemade, (2013), Study of Pond Ash (BTPS) Use as A Fine Aggregate in Cement Concrete - Case Study, International Journal of Latest Trends in Engineering and Technology (IJLTET) Vol. 2, pp. 292-297.
26. M. Romeekadevi and K. Tamilmullai. (2015), Effective Utilization of Fly Ash and Pond Ash in High Strength Concrete, International Journal of Engineering Research & Technology (IJERT)-special issue, Vol. 3, pp.1-7.
27. B. Ganesh, H.S. Bai, R. Nagendra, and S. Bagad, (2011), Pond Ash: An Alternative Material as Fine Aggregate in Concrete for Sustainable Construction, Advanced Materials Research, Vol. 306-307, pp. 1071-1075 (DOI:10.4028/www.scientific.net/AMR.306-307.1071).
28. A. Ullah, A. Kassim, A. Abbil, S Matusin, A.S.A Rashid, N.Z.M. Yunus, and R Abuelgasim, (2020), Evaluation of Coal Bottom Ash Properties and Its Applicability as Engineering Material, IOP Conf. Series: Earth and Environmental Science, Vol. 498, pp. 1-8 (DOI:10.1088/1755-1315/498/1/0120441).
29. A. Bhatta, S. Priyadarshinia, A.A. Mohanakrishnana, A. Abria, M. Sattlera, and S. Techapaphawitc, (2019), Physical, Chemical, and Geotechnical Properties of Coal Fly Ash: A Global Review, Case Studies in Construction Materials, Vol. 11, pp.1-11 (https://doi.org/10.1016/j.cscm.2019.e00263).
30. E. Haustein and A. Kuryłowicz-Cudowska, (2020), The Effect of Fly Ash Microspheres on the Pore Structure of Concrete, Minerals, Vol.10, pp. 1-12 (doi:10.3390/min10010058).
31. A. Sarkara, R. Ranoa, K.K. Mishra, and I.N. Sinha, (2005), Particle Size Distribution Profile of Some Indian Fly Ash-A Comparative Study to Assess Their Possible Uses, Fuel Processing Technology, Vol. 86, pp. 1221-1238 (DOI:10.1016/j.fuproc.2004.12.002).
32. F. Wu, H. Li, and K. Yang, (2021), Effects of Mechanical Activation on Physical and Chemical Characteristics of Coal-Gasification Slag, Coatings, Vol. 11, pp. 1-15 (https://doi.org/10.3390/coatings11080902).
33. M. Matsuoka, K. Yokoyama, K. Okura, N. Murayama, M. Ueda, and M. Naito, (2019), Synthesis of Geopolymers from Mechanically Activated Coal Fly Ash and Improvement of Their Mechanical Properties, Minerals, Vol. 9, pp. 1-10 (DOI:10.3390/min9120791).
34. S. Rios, N. Cristelo, T. Miranda, N. Araújo, J. Oliveira and E. Lucas, (2016), Increasing the Reaction Kinetics of Alkali-Activated Fly Ash Binders for Stabilization of A Silty Sand Pavement Sub-Base, Road Materials and Pavement Design, Vol. xxx, pp.1-23 (DOI:10.1080/14680629.2016.1251959).
35. P. Rozek, P. Florek , M. Król and W. Mozgawa, (2021), Immobilization of Heavy Metals in Boroaluminosilicate Geopolymers, Materials, Vol. 14, pp. 1-16 (https://doi.org/10.3390/ma14010214).
36. V. Nikolic, M. Komljenović, N. Marjanovic, and Z. Bascarevic (2014), Lead immobilization by geopolymers based on mechanically activated fly ash, Ceramics International, Vol. 40, pp. 8479-8488 (DOI:10.1016/j.ceramint.2014.01.059).

37. R. S. abolaji, Z. Bingxue, R. Guarecuco, and T. Thomas, (2018), Geopolymer for use in heavy metals adsorption, and advanced oxidative processes: A critical review, Journal of Cleaner Production, Vol. 213, pp.xxx-xxx (DOI:10.1016/j.jclepro.2018.12.145).
38. M. Alshaaera, M. Shqaira, H.G. Abdelwaheda, K. Abuhaseld, M.Z. Toro, (2017), Stabilization of heavy oil fly ash (HFO) for construction and environmental purposes, International Journal of Applied Engineering Research, Vol. 12, pp. 488-497.
39. B. Kim, B. Lee, C.-M. Chon, and S. Lee, (2019), Compressive Strength Properties of Geopolymers from Pond Ash and Possibility of Utilization as Synthetic Basalt, Journal of the Korean Ceramic Society, Vol. 56, pp. 365-373. (https://doi.org/10.4191/kcers.2019.56.4.03)
40. S.K. Saxena, N.B. Singh, and M. Kumar, (2017), Influence of Alkali Solutions on Properties of Pond Fly Ash-based Geopolymer Mortar Cured under Different Conditions, Advances in Cement Research, Vol. xxx, pp. 1-7 (http://dx.doi.org/10.1680/jadcr.17.00038).
41. M. Khodr, D.W. Law, C. Gunasekara, S. Setunge and R. Brkljaca, (2019), Compressive Strength and Microstructure Evolution of Low Calcium Brown Coal Fly Ash-based Geopolymer, Journal of Sustainable Cement-Based Materials, Vol. xxx, pp. 1-19 (DOI: 10.1080/21650373.2019.1666061).
42. M. Panigrahi, S. Mohanty, R.R. Dash, R.I. Ganguly, (2018) Development of Novel Constructional Material from Industrial Solid Waste as Geopolymer, IOP Conf. Series: Materials Science and Engineering, Vol. 410, pp. 1-12 (DOI:10.1088/1757-899X/410/1/012012).
43. M.P. Bhamare, Y.N. Bafna, and A.K. Diwedi, (2012), Engineering Properties of Cement Containing Pond Ash, ISRO Journal of Engineering, Vol. 2, pp. 7-11.
44. M. Panigrahi, P. Sivakrishnan, P.K. Ran, R.R. Dash, R.I. Ganguly, (2016), Structural Transformation of Pond Ash Geopolymer: A Novel Construction Material" in 20th National Conference on Non-Ferrous Minerals and Metals-2016.
45. FTIR Spectroscopy Attenuated Total Reflectance (ATR), www. perkinelmer. com.
46. W.E. Smith and G. Dent, (2005), Modern Raman Spectroscopy-A Practical Approach, John Wiley & Sons, Ltd, (ISBNs: 0-471-49668-5 (HB); 0-471-49794-0 (PB)).
47. M. Alexandre, and P. Dubois, (2000), Polymer-layered Silicate Nanocomposites: Preparation, Properties and Uses of a New Class of Materials, Materials Science and Engineering R, Vol. 28, pp. 1-63.
48. H.P. Klugg and L.E. Alexander, (1974), X-ray Diffraction Procedures; John Wiley & Sons: U.S.A.
59. D. Moore, and R. Jr. Reynolds, (1997), X-ray Diffraction and Identification and Analysis of Clay Minerals of Clay Minerals; Second Ed.; Oxford University Press, U.K.

50. Carl Zeiss Microscopy, (2004), Detection Principles based on GEMINI Technology.
51. Anandhan, S. Thermal Analysis, Dept. of Met. and Mat. Engg., NITK.
52. P. Gabbott (Editor), (2008), Principles and Applications of Thermal Analysis, John Wiley & Sons, Inc., pp. 1999-2013 (DOI: 10.1002/9780470697702).
53. Alehyen, S., Achouri, M. El, and Taibi, M., (2017), Characterization, Microstructure and Properties of Fly Ash-based Geopolymer, Journal of Materials and Environmental Sciences (JMES), Vol. 8, pp. 1783-1796.
54 A.M. Mustafa Al Bakri, H. Kamarudin, M. Bnhussain, I.N. Khairul, A.R. Rafiza, and Y. Zarina, (2012), The Processing, Characterization, And Properties of Fly Ash Based Geopolymer Concrete, Reviews on Advanced Materials Science, Vol. 30 pp. 90-9.
55 F. Škvára, L. Kopecky, V. Šmilauer, and Z. Bittnar, (2009), Material and Structural Characterization of Alkali Activated Low-Calcium Brown Coal Fly Ash. J Hazard Mater, Vol. 168, pp.711-20.
56. Chindaprasirt P, Thaiwitcharoen S, Kaewpirom S, Rattanasak U. (2013), Controlling Ettringite Formation in FBC Fly Ash Geopolymer Concrete, Cem Concr Compos, Vol. 41, pp. 24-8.
57. P.N. Lemougna, K.J.D. MacKenzie, and U.F.C. Melo. (2011), Synthesis and Thermal Properties of Inorganic Polymers (Geopolymers) for Structural and Refractory Applications from Volcanic Ash. Ceram Int, Vol. 37, pp. 3011-8.
58. F. Winnefeld, A. Leemann, M. Lucuk, P. Svoboda, and M. Neuroth, (2010), Assessment of phase formation in alkali activated low and high calcium fly ashes in building materials. Constr Build Mater, Vol. 24, pp.1086–93.
59. P. Chindaprasirt and U. Rattanasak, (2010), Utilization of Blended Fluidized Bed Combustion (FBC) Ash and Pulverized Coal Combustion (PCC) Fly Ash in Geopolymer, Waste Manage, Vol. 30, pp. 667-72.
60. J.G.S. Van Jaarsveld, J.S.J. Van Deventer, and G.C. Lukey, (2003), The Characterization of Source Materials in Fly Ash-based Geopolymers, Materials Letters, Vol. 57, pp. 1272-1280.
61. C. Jeyageetha, S.P. Kumar, (2016), Study of SEM/EDXS and FTIR for Fly Ash to Determine the Chemical Changes of Ash in Marine Environment, International Journal of Science and Research (IJSR), Vol. 5, pp. 1688-1693.
62. M. Criado, A. Fernández-Jiménez, and A. Palomo, (2007), Alkali Activation of Fly Ash: Effect of the SiO_2/Na_2O Ratio Part I: FTIR Study, Microporous and Mesoporous Materials, Vol. 106, pp. 180-191 (DOI:10.1016/j.micromeso.2007.02.055).
63. I. Wilinska and B. Pacewska, (2019), Comparative Investigation of Reactivity of Different Kinds of Fly Ash in Alkaline Media, Journal of Thermal Analysis and Calorimetry, Vol.138, pp. 3857-3872 (https://doi.org/10.1007/s10973-019-08296-4).

64. A.M. Kalinkin, B.I. Gurevich, M.S. Myshenkov, M.V. Chislov, E.V. Kalinkina, I.A. Zvereva, Z. Cherkezova-Zheleva, D. Paneva and V. Petkova, (2020), Synthesis of Fly Ash-Based Geopolymers: Effect of Calcite Addition and Mechanical Activation, Minerals, Vol. 10, pp.1-21 (DOI:10.3390/min10090827)
63. Palacios, M., Banfill, P.F.G., and Puertas, F., (2018), Rheology and setting of alkali-activated slag pastes and mortars: Effect of organ admixture, *ACI Materials Journal*, Vol.xxx (Title no. 105-M16), pp.140-148.
64. Chindaprasirt, P., Chareerat, T., and Sirivivatnanon, V., (2007), Workability and strength of coarse high calcium fly ash geopolymer, cement and concrete composites, Vol. 29, pp. 224-229.
65. Zivica, V., Palou, M., and Ifka Ľ. Bágeľ, T., (2014), High strength metahalloysite based geopolymer, Composites Part B Engineering, Vol. 57, pp. 155-165, DOI: 10.1016/j.compositesb.2013.09.034.
66. AS 1012.9:2014 is Methods of testing concrete, Method 9: Compressive strength tests — Concrete, mortar and grout specimens This Standard sets out the method for determining the compressive strength of concrete test specimens prepared in accordance with the provisions of AS 1012.8.1, AS 1012.8.3, AS 1012.14 or AS 1012.19 (Published: 27/06/2014).

4

Quantification of Variables on Strength Property of Pond Ash (PA)-Based Geopolymer

Muktikanta Panigrahi[1]*, Subhasmita Prusty[2], Ratan Indu Ganguly[3] and Radha Raman Dash[4]

[1]PG Department of Materials Science, Maharaja Sriram Chandra Bhanja Deo University, Keonjhar Campus, Odisha, India
[2]Department of Civil Engineering, Gandhi Institute of Engineering and Technology University, Gunupur, Odisha, India
[3]Department of Metallurgical Engineering, National Institute of Technology, Raurkela, Odisha, India
[4]CSIR-National Metallurgical Laboratory, Jamshedpur, Jharkhand, India

Abstract

Concrete is the world's most versatile, durable and reliable building material in the construction industry. One of the major ingredients of concrete is ordinary Portland cement (OPC). It is the second major generator of carbon dioxide, causing pollution in the atmosphere. The process is energy-intensive. Hence, it was inevitable that an alternative material would be considered for cost-effective and reliable construction applications. Geopolymer concrete is an innovative construction material which can be produced by chemical activation of inorganic molecules such as silica and alumina bearing phases which are found in raw materials (fly ash or pond ash) used for making geopolymers. In this chapter, quantification of variables has been made relating to strength properties of geopolymer. For quantification, a statistical design of an experiment is planned, a regression equation developed, and the regression equation has been statistically verified to find the accuracy of the equation. Prediction of geopolymer strength properties in

**Corresponding author*: muktikanta2@gmail.com

Muktikanta Panigrahi, Ratan Indu Ganguly and Radha Raman Dash (eds.) Development of Geopolymer from Pond Ash-Thermal Power Plant Waste: Novel Constructional Materials for Civil Engineers, (123–150)

a range of variations and variables are studied. The presence of significant interaction coefficients in regression equations indicates that equations are nonlinear in nature, suggesting adoption of higher order design for a more accurate prediction of mechanical properties. Analysis by XRD, FTIR, FESEM with EDS, DSC and TGA of GP prepared from optimum parameters is also carried out.

Keywords: Geopolymer, mechanical strength, statistical design, water crystallization, thermal stability

4.1 Introduction

Because the use of waste and by-products in geopolymer concrete will lead to a greener environment, it can be called "green concrete." Different types of waste materials can be considered for use in geopolymer concrete. The most commonly used waste materials to replace sand and cement in concrete are fly ash, rice husk ash, blast furnace slag, red mud, phosphor, gypsum, silica fume, fumed silica, crushed glass, and egg shells. Ash is the residue after combustion of coal in thermal power plants. Particle sizes of the ash vary from one micron to 600 microns. Very fine fly ash particles collected from this ash generated in electrostatic precipitators are being used in the manufacture of blended cements. Unused fly ash and bottom ash (residue collected at the bottom of a furnace) mixed in slurry form and deposited in ponds are known as pond ash. Thermal power plants in India adopt wet methods of disposal and storage of ash in large ponds and dykes. In the wet method, both fly ashes collected from electrostatic precipitators and bottom/grate ash are mixed with water and transported to the ponds in a slurry form. Chemical and mineralogical properties of fly ash depend on the characteristics and composition of coal fed into the boiler furnace. Basically, it contains large amounts of silica (SiO_2), alumina (Al_2O_3), and iron oxide (Fe_2O_3); and small quantities of other oxides like CaO, MgO, MnO, Na_2O, K_2O, SO_3 in the form of minerals, *viz.*, magnetite, hematite, quartz, free calcium oxide, etc. Disposal of fly ash and pond ash will be a big challenge to the environment, especially when the quantum increases from the present level. Hence, worldwide research is focused on finding an alternative use for this waste material. Its use in the concrete industry is one of the effective methods of utilization.

Bottom ash is a by-product of coal burnt in thermal power plants. Bottom ash particles are much coarser than fly ash. It is a coarse, angular material with a porous surface. This material is composed of silica, alumina, and iron oxide with small amounts of calcium, magnesium and sulphate. Grain size typically ranges between finer sand size to coarser gravel size. The chemical composition of bottom ash is similar to fly ash but typically contains greater amounts of carbon. Bottom ash exhibits high shear strength and low compressibility. These engineering properties make bottom ash an ideal material for the design and construction of dams and for other civil engineering applications. Bottom ash also exhibits a relatively high permeability and grain size distribution that allows the design engineer to use it in direct contact with impervious material. Bottom ash has proved to be an economical material because it has been demonstrated to not only have good engineering property but also to have constructability benefits. Bottom ash can be used as concrete aggregate or for several other civil engineering applications where sand, gravel and crushed stone are used. Geopolymer is the new generation construction material proposed to replace conventional materials such as cement. Several workers have developed geopolymer from fly ash. These materials are found to be a better replacement for cement due to their ease of production and cost-effectiveness; and yet show better performance in terms of strength, stability and structural integrity.

Geopolymer also saves energy and causes less environmental pollution. The process of producing geopolymer also uses industrial waste such as fly ash, pond ash, etc. However, fly ash is no longer considered a waste product because it is being marketed to industries dealing with other construction materials such as bricks and pavements [1, 2].

Geopolymer can also be produced by using raw materials such as pond ash, slag or steel melting slag. It is prepared by varying parameters such as size of ingredients, raw materials, curing temperature, and curing time. These parameters need to be optimized by trial-and-error method and varying the parameters one at a time (single factor experiment).

In single factor experiment, one parameter can vary, while others are kept constant. Thus, to get the optimization values of three parameters many experiments are needed, and yet optimization still may not be possible. This is because the interactions between the parameters are ignored at their various levels. A statistical design experiment solves this problem by simultaneously varying all the operating parameters within some range [3–7]. For that, factorial design is adopted using no. of experiments "P^n";

where P = level of variation and n = no. of variable.

In this case, the level of variation is two, i.e., p = 2, and no. of variables are three, i.e., n = 3. Therefore, if factors are varied at two levels than

8 no. of experiments are required for construction design matrix, which will yield response surface connecting properties such as strength with variables. This equation will give a quantity estimate of the materials with respect to the response such as strength.

If the range of variation is very narrow then the response surface may be an approximate plane. In such case, the regression equation so developed will resolve more production variables in terms of main effect [3–7]. However, the equation needs to be statically refined by a T-test and be further tested for adequacy by an F-test.

In the present investigation, pond ash is taken as raw material with alkaline compounds of the laboratory used in a fixed ratio.

4.2 Experimental Details

4.2.1 Materials and Method

Pond Ash was collected from NALCO Navratna Company (Damanjodi, Odisha). The chemical composition/different phases of pond ash are shown in Table 4.1. Sodium hydroxide (NaOH) was purchased from Loba Chemicals. Sodium silicate (Na_2SiO_3) was purchased from Merck, India. A few drops of water-soluble plasticizer (Sika) was used to prepare geopolymer in order to improve the plasticity of as-prepared materials. A small amount of water was used to mix the raw materials.

4.2.2 Preparation of Geopolymer from Raw Materials

The geopolymer process was carried out in three steps, i.e., grinding, mixing and ramming, in the steps shown below [8, 9]:

Step 1: Pond ash is ground by mortar and pestle and then screen through 240 meshes.
Step 2: Sieved pond ash is mixed with NaOH, Na_2SiO_3, a small amount of water (6%) and water-soluble plasticizers (Sika). Gel-like material is formed. Mixtures of different compositions are shown in Table 4.2.
Step 3: Gel-like materials are put in REMI system. The REMI mold is made of iron having cylindrical shaped (50 mm diameter by 70 mm height). The samples are rammed twenty (20) times, compacted for 3 minutes and then demolded. These are called green samples. Green samples are cured at 70 °C for 24 h in an air oven. The cured samples are preserved in plastic zipper bags to avoid moisture loss. The samples are tested.

4.2.3 Characterization of Prepared Samples

Spectroscopic studies were conducted using a field emission scanning electron microscope (FESEM) with an energy-dispersive X-ray spectrometer (JEOL). Before examination, gold coating was done using the sputter technique.

Compression testing of as-prepared geopolymer samples was performed in an automatic compression testing machine (AIMIL COMPTEST 2000, India), followed by ASTM standard procedure.

Analysis of XRD samples were done using a Phillips PW-1710 advance wide angle X-ray diffractometer, Phillips PW-1729 X-ray generator and Cu Kα radiation (wavelength, $\lambda = 0.154$ nm).

The FTIR spectra were recorded on a Thermo Nicolet Nexus 870 spectrometer in the range of 400-4000 cm^{-1}.

Thermal stability was determined by TGA; and the water of crystallization was calculated using DSC (Perkin Elmer Pyris Diamond calorimeter) at a heating rate of 10 °C/min in nitrogen environment.

For optimization purpose, judicious selection of input variables at base level is required. Activator solution and three variables, such as curing time, curing temperature and mesh sizes, are selected to evaluate the accuracy of the ACI 211.1 practice to predict the compressive strength of PA-based geopolymer. Two design variables are included to analyze the results at three different levels of compressive strength. Also, five factors (control parameters) are kept constant throughout the experimental plan.

A design of experiments for the activator solution control variables for each combination of the two variables is created using Minitab software, which provided the experiments required and randomized order of testing. It also provided the analysis of variance of the results and the deviation from the expected outcome, providing data that was fed into the mix design and proportionating software.

4.3 Results and Discussion

There is a lack of information regarding the quantitative effect of variables on strength properties of geopolymer. The numbers of processing variables which affect strength properties are many, and consequently complex interactions result due to these variables interacting with each other. To reach an optimum level of variables for maximum properties, one has to perform a large number of experiments, and even then, the optimum combination is not always guaranteed. Under this circumstance use of statistical design of experiments is extremely helpful. By performing fewer experiments in a planned manner, one can reach the optimum combination of variables in much shorter time, thus saving considerable labor and cost. This mathematical tool has been used profitably by two of the authors. In the present situation, the utility of statistical design of experiments is once again demonstrated for geopolymer prepared with pond ash.

Many factors affected the mechanical properties of geopolymer during the present investigation; the following factors are varied:

1) Curing temperature (x1)
2) Particle size (x2)
3) Curing time (x3)

Various workers suggested artificial curing temperature and time of the geopolymers in the range of 50–90 °C and 20–28 h, respectively, for obtaining optimum properties. Based on this information and the single factor experiments performed by the investigators, base levels for curing time and temperature were decided. Since curing time, curing temperature and particle size interact in a complex manner, it was decided to develop a regression equation connecting the input variables like curing temperature, particle size and curing time with the response, i.e., strength properties. For working out the optimum treatment combination from this regression equation, experiments were performed by following the method of steepest ascent (77-78). The optimum treatment combination thus obtained was employed for treatment of geopolymer. Regression equations developed between the response variables and the factors varied. Table 4.1 lists the range of variations of the different factors and base level chosen together with the code for each. Table 4.2 shows the composition of three base levels of variables from which geopolymer is prepared. However, all the analyses were repeated to see the reproducibility.

In order to establish the quantitative relationship between residual compressive strength and composition of the geopolymer, statistical design was applied. Here, each geopolymer element was varied in two levels, keeping the base level geopolymer composition as the central composition. Table 4.2 shows variation of input variables in two levels (Ref. Table 4.1). Since there are variables varying between two levels, a design matrix is constructed by simultaneous variation of three factors. Hence, eight combinations are required following the factorial design experiment, where a

Table 4.1 Variation of different factors (geopolymers), base levels and codes.

Factors varied code	**Temperature (x_1)**	**Mesh size (x_2)**	**Curing time (x_3)**
Upper level	90	240	28
Code value	(+1)	(+1)	(+1)
Base level	70	195	24
Code value	(0)	(0)	(0)
Lower level	50	150	20
Code value	(-1)	(-1)	(-1)

Table 4.2 Composition of eight geopolymers prepared with their code values and residual compressive strength.

Sl. no.	**Input variables**			**Residual compressive strength (MPa)**
1	90 (+1)	240 (+1)	28 (+1)	18.11
2	50 (-1)	240 (+1)	28 (+1)	15.10
3	90 (+1)	150 (-1)	28 (+1)	15.10
4	50 (-1)	150 (-1)	28 (+1)	14.22
5	90 (+1)	240 (+1)	20 (-1)	20.02
6	50 (-1)	240 (+1)	20 (-1)	12.22
7	90 (+1)	150 (-1)	20 (-1)	16.88
8	50 (-1)	150 (-1)	20 (-1)	10.54

matrix is constructed, i.e., P^n, where P represents level of variation and n represents number of factors. Thus, $2^3 = 8$ experiments are required for three variables for constructing a design matrix. For each treatment, the combinations response, i.e., strength properties, are shown in Table 4.2.

The response obtained in Table 4.2 was analyzed and regression coefficient was obtained. Table 4.3 shows regression coefficients obtained for different treatment combinations.

There are three variables, namely curing temperature (x1), particle size (x2) and curing time (x3), and hence a number of experiments were required in the design matrix. Number of experiments is 2^3, i.e., 8 treatment combinations, is shown in Table 4.2. Table 4.2 shows the composition and coded values of eight geopolymers prepared, together with response obtained for each geopolymer.

The regression equation obtained was formed using data from Table 4.2:

$$Y = b_o \pm b_1x_1 \pm b_2x_2 \pm b_3x_3 \pm b_{12}x_1x_2 \pm b_{23}x_2x_3 \pm b_{13}x_1x_3 + b_{123}x_1x_2x_3$$

Calculation of regression coefficients

Where, $b_o = (18.11 + 15.10 + 15.10 + 14.22 + 20.02 + 12.22 + 16.88 + 10.54) \div 8$
$= 15.27$

$,b_1 = ((18.11 + 15.10 + 20.02 + 16.88) - (15.10 + 14.22 + 12.22 + 10.54)) \div 8$
$= 2.25$

Table 4.3 Regression coefficients calculated from the response values obtained for different treatment combinations.

Regression coefficient	**Compressive strength of geopolymer**
b0	15.27
b1	2.25
b2	1.08
b3	0.36
b12	0.44
b13	-1.28
b23	-0.11
b123	0.083

$$,b_2 = ((18.11 + 15.10 + 20.02 + 12.22)\text{-}(15.10 + 14.22 + 16.88 + 10.54)) \div 8 = 1.08$$

$$,b_3 = ((18.11 + 15.10 + 15.10 + 14.22)\text{-}(20.02 + 12.22 + 16.88 + 10.54)) \div 8 = 0.36$$

$$,b_{12} = ((18.11 + 14.22 + 20.02 + 10.54)\text{-}(15.10 + 15.10 + 12.22 + 16.88)) \div 8 = 0.44$$

$$,b_{23} = ((18.11 + 15.10 + 16.88 + 10.54)\text{-}(15.10 + 14.22 + 20.02 + 12.22)) \div 8 = -0.11$$

$$,b_{13} = ((18.11 + 15.10 + 12.22 + 10.54)\text{-}(15.10 + 14.22 + 20.02 + 16.88)) \div 8 = -1.28$$

$$,b_{123} = ((18.11 + 14.22 + 12.22 + 16.88)\text{-}(15.10 + 15.10 + 20.02 + 10.54)) \div 8 = 0.083$$

So, the equation becomes,

$$Y = 15.27 + 2.25x_1 + 1.08x_2 + 0.36x_3 + 0.44x_1x_2 \text{-} 0.11x_2x_3 \text{-} 1.28x_1x_3 + 0.083x_1x_2x_3,$$

$$X_1 = \frac{x1 - 70°c}{20°c}$$

$$X_2 = \frac{x2 - 195}{45}$$

$$X_3 = \frac{x3 - 24}{4}$$

Where, x1, x2 and x3 are the main effect of curing temperature, particle size and curing time varying between ±1. In order to calculate the residual compressive strength value for any composition of the geopolymer in the range of variation of the geopolymer elements, code values are to be inserted in the above equation instead of using the actual composition in percent. The validity of the equation is assured only within the range of composition studied in the investigation.

The equation is nonlinear in nature and there are several binary and ternary coefficients (coefficient attached to x1x2 or x1x2x3 in Eq. 4.1). It was decided to plan experiments by constructing a separate design matrix for finding the exact combination of curing temperature, mesh size curing, and time. Thus, the variables are curing temperature, aging time, and particle size. The base level of treatment combination was decided after the single factor experiment. Thus, the base level for curing temperature, curing time, and particle size is 70°, 240, 24 h respectively. Table 4.4 shows the design matrix with different values obtained for each experiment.

A few coefficients might be become statistically insignificant for the following reasons:

1) Variation from batch to batch;
2) Error in measurement of different response variables.

The latter may be subdivided under three heads:

a) Variation from sample to sample within the same batch;
b) Due to equipment;
c) Personal error while preparing geopolymer, chemical analysis and heat treating the samples, etc.

Table 4.4 Experimental design matrix showing treatment combinations.

	Input variables			
Sl. no. code values	**Temp. (x1)**	**Mesh size (x2)**	**Curing time (x3)**	**Treat. comb.**
1 (Code value)	90 (+1)	240 (+1)	28 (+1)	x1x2x3
2 (Code value)	50 (-1)	240 (+1)	28 (+1)	x2x3
3 (Code value)	90 (+1)	150 (-1)	28 (+1)	x1x3
4 (Code value)	50 (-1)	150 (-1)	28 (+1)	x3
5 (Code value)	90 (+1)	150 (+1)	20 (-1)	x1x2
6 (Code value)	50 (-1)	240 (+1)	20 (-1)	x2
7 (Code value)	90 (+1)	150 (-1)	20 (-1)	x1
8 (Code value)	50 (-1)	150 (-1)	20 (-1)	1

All these errors are therefore responsible for making values of some coefficients statistically insignificant. However, to find significant coefficients for corresponding effects and interactions, analysis of variance is required.

4.3.1 Testing of Significance Coefficients

Significance of each coefficient is tested separately. This can be done in two equivalents ways, namely, by using "student's t" or by constructing the confidence interval. When a complete factorial experiment or a regular fractional replication is used, the confidence intervals for all the coefficients (including the interaction effects) are equal to one another.

First of all, it is necessary, of course, to find the variance of the regression coefficients $S^2(bj)$;

$$S^2\,bj = \frac{s2y}{N}$$

Where, N=Total number of experiments

$$Sy^2 = \frac{\sum_1^N \sum_1^n (y0-y)2}{N(n-1)}$$

Where, Y_0= Response obtained for each replicate
Y= Average of response for n replicate
n= Number of replicate made at each level

A glance at the formula shows that variances of all the coefficients are equal to one another, since they depend only on the error of the experiment and the number of trials. Now it is easy to construct the confidence interval (Δ bj)

$$(\Delta\,bj) = \pm ts(bj)$$

Here, 't' is tabulated value of 'student's' criterion for the number of degrees of freedom $S^2(y)$, which is determined with and at the selected significance level (usually 0.05 or 5%), s(bj) is the quadratic error of the regression coefficient.

$$S(bj) = +\sqrt{s2(bj)}$$

The formula for the confidence interval can be written in the following equivalent form:

$$bj = \pm \frac{ts(y)}{\sqrt{N}}$$

A coefficient is significant if its absolute value is greater than confidence interval.

The confidence interval is set up by the upper and lower limits bj+Δbj and bj-Δbj.

The value of t can be found from Table 4.5, a fragment of which is show in Table 4.6, where f is the number of degrees of freedom.

From the above data the different Δbj values are estimated.

Table 4.5 Student's t at 5% significance level (12,61).

f	t	f	t	f	t
1	12.71	13	2.160	26	2.056
2	4.303	14	2.145	27	2,052
3	3.182	15	2.131	28	2.048
4	2.776	16	2.120	29	2.045
5	2.571	17	2.110	30	2.042
6	2.447	18 19	2.101 2.093		1.960
7	2.365	20	2.086		
8	2.306	21	2.080		
9	2.262	22	2.074		
10	2.228	23	2.069		
11	2.201	24	2.064		
12	2.179	25	2.060		

Table 4.6 Computation of Δbj for compressive strength values.

Treat comb.	Y0	Y	(Y0-Y)	(Y0-Y)2
x1x2x3	19 18 17.33	18.11	0.89 -0.11 -0.78	0.792 0.012 0.608
x2x3	14.66 15 15.66	15.10	-0.44 -0.1 0.56	0.193 0.01 0.313
x1x3	14.66 15.66 15	15.10	-0.44 0.56 -0.1	0.193 0.313 0.01
x3	14 14.66 14	14.22	-0.22 0.44 -0.22	0.048 0.193 0.048
x1x2	20.03 20.03 20	20.02	0.01 0.01 -0.02	0.0001 0.0001 0.0004
x2	12.34 12.33 12.0	12.22	0.12 0.11 -0.22	0.014 0.012 0.048
x1	17 17 16.66	16.88	0.12 0.12 -0.22	0.014 0.014 0.048
1	10.33 10.0 11.3	10.54	-0.21 -0.54 0.76	0.044 0.291 0.577

Any coefficients of Equations 4.5 to 4.8 having a value less than Δbj value are to be ignored, then new equations are formed after ignoring the insignificant coefficients.

$$\Sigma(y0-y)2 = 3.795$$

$$S^2y = \frac{\Sigma_1^N \Sigma_1^n (y0-y)2}{N(n-1)} = \frac{\Sigma_1^8 \Sigma_1^3 (y0-y)2}{8(3-1)} = \frac{3.795}{8(3-1)} = \frac{3.795}{16} = 0.237$$

$$S^2\ bj = \frac{s2y}{N} = \frac{0.237}{8} = 0.029 \qquad \text{Eq (4.1)}$$

$$Sbj = \sqrt{0.029} = 0.170$$

$$\Delta\ bj = ta, N\ sbj = 2.306{*}0.70 = 0.392$$

The master equation is,

$$Y = 15.27 + 2.25x1 + 1.08x2 - 1.28x1x3$$

$$Sad^2 = \frac{\Sigma\Delta y2}{f} = \frac{2.803}{4} = 0.701$$

$$F = N\text{-}k$$

$$= 8\text{-}4$$

$$= 4$$

The master equation is,

$$Y = 15.27 + 2.25x_1 + 1.08x_2 - 1.28x_1x_3$$

The adequacy of the equation is checked by variance ratio test. For this purpose, we have to calculate the residual variance or the variance of adequacy (s^2 ad), which is given by,

$$s2\,ad = \frac{\Sigma_{i=1}^N \Delta Yi2}{f} \qquad \text{Eq (4.2)}$$

Where, Δyi = Difference between the experimental and calculated properties.

f = Number of degree

which is denoted by

$$f = (N\text{-}K)$$

Where, N = Total number of experiments
K = Number of significant coefficients of regression equation.

The value of F is calculated by using the formula,

$$\text{Fcal} = \frac{s2\,ad}{s2\,(y)} \qquad \text{Eq (4.3)}$$

Where s^2 ad is given by Eq. (4.2) and s^2 (y) is given by Eq. (4.1) respectively.

If Fcal is less than the value of F obtained from standard statistical table, then the model is adequate and the equation may be used to predict the property correctly within the range of variation of variables.

Table 4.7 shows calculated (using Eq. 4.3) and experimentally obtained values. Since number of significant coefficients varied for different equations, the value of 'f' also differs.

Values of 'F' are calculated by using the formula and 'F' obtained from the standard F-distribution table, for the required degree of freedom mentioned at the matrix.

Since Fcal≤ *Ftable*, the equations are adequate in predicting the mechanical properties of the geopolymer at 5% point in the range of variation

Table 4.7 Computation of "F" value for compressive strength.

Matrix	Experimental	Calculated	Δy	$\Delta y2$
x1x2x3	18.11	17.32	0.79	0.624
x2x3	15.10	15.38	-0.28	0.078
x1x3	15.10	15.16	-0.06	0.0036
x3	14.22	13.22	1.0	1.0
x1x2	20.02	19.88	0.14	0.019
x2	12.22	12.82	-0.6	0.36
x1	16.88	17.72	-0.84	0.705
1	10.54	10.66	-0.12	0.014

studied in the present investigation. Therefore, the equation is accurate in the range of variation of variables.

Accuracy of the equation is further checked by random experiments. For this, value of response variables for base level geopolymer is accepted to the put to zero and b0 calculated from the equation and is compared with those obtained by actual experiments.

$$\text{Now, } \mathbf{F} = \frac{Sad2}{s2y} = \frac{0.701}{0.237} = \mathbf{2.95}$$

$$\mathbf{F_{4,\,8} = 3.8}$$

Hear we found, Fcal = 2.95 and Ftable = 3.8

So that, **Fcal≤Ftable**

The response variables obtained for base level geopolymer being treated at the base level of time and temperature of aging very closely match the 'b_o' values obtained by regression analysis. The 'b_o' values correspond to the properties of base level geopolymer being aged at base level of time and temperature. This also provides a picture of probable difference in mechanical properties that we might obtain in the measured and calculated value of mechanical properties of any other random geopolymer chosen in the range of composition studied.

The experimental and calculated values closely match, though some variation is observed. This is attributed to the non-linearity of the regression equations and also for variation in composition of the base level geopolymer. For a more accurate prediction of properties, it is desirable to adopt higher order design.

These equations can be used to design geopolymer with specific properties. A computer may be used to find the combination of the input variables for obtaining maximum strength properties in the geopolymer. Here, of course, constraints should be imposed on the range of variation of the variables and ductility of the geopolymer. If such equations are also available for some other properties, like corrosion and impact properties, they may also be used as constraints when maximizing the strength properties.

Table 4.8 Values of F ratio at 5% significance level.

Sl. no.	f1/f2	1	2	3	4	5	6	12	24	
1	1	164.4	199.5	215.7	224.6	230.2	234.0	244.9	249.0	254.3
2	2	18.5	19.2	19.2	19.3	19.3	19.3	19.4	19.4	19.5
3	3	10.1	9.6	9.3	9.1	9.0	8.9	8.7	8.6	8.5
4	4	7.7	6.9	6.6	6.4	6.3	6.2	5.9	5.8	5.6
5	5	6.6	5.8	5.4	5.2	5.1	5.0	4.7	4.5	4.4
6	6	6.0	5.1	4.8	4.5	4.4	4.3	4.0	3.8	3.7
7	7	5.5	4.7	4.4	4.1	4.0	3.9	3.6	3.4	3.2
8	8	5.3	4.5	4.1	3.8	3.7	3.6	3.3	3.1	2.9
9	9	5.1	4.3	3.9	3.6	3.5	3.4	3.1	2.9	2.7
10	10	5.0	4.1	3.7	3.5	3.3	3.2	2.9	2.7	2.5
11	11	4.8	4.0	3.6	3.4	3.2	3.1	2.8	2.6	2.4
12	12	4.8	3.9	3.5	3.3	3.1	3.0	2.7	2.5	2.3
13	13	4.7	3.8	3.4	3.2	3.0	2.9	2.6	2.4	2.2
14	14	4.6	3.7	3.3	3.1	3.0	2.9	2.5	2.3	2.1

(Continued)

Table 4.8 Values of F ratio at 5% significance level. (*Continued*)

Sl. no.	f1/f2	1	2	3	4	5	6	12	24	
15	15	4.5	3.7	3.3	3.1	2.9	2.8	2.5	2.3	2.1
16	16	4.5	3.6	3.2	3.0	2.9	2.7	2.4	2.2	2.0
17	17	4.5	3.6	3.2	3.0	2.8	2.7	2.4	2.2	2.0
18	18	4.4	3.6	3.2	2.9	2.8	2.7	2.3	2.1	1.9
19	19	4.4	3.5	3.1	2.9	2.7	2.6	2.3	2.1	1.8
20	20	4.4	3.5	3.1	2.9	2.7	2.6	2.3	2.1	1.8
21	22	4.3	3.4	3.1	2.8	2.7	2.6	2.2	2.0	1.8
22	24	4.3	3.4	3.0	2.8	2.7	2.6	2.2	2.0	1.8
23	26	4.2	3.4	3.0	2.7	2.6	2.5	2.2	2.0	1.7
24	28	4.2	3.3	3.0	2.7	2.6	2.4	2.1	1.9	1.6
25	30	4.2	3.3	2.9	2.7	2.5	2.4	2.1	1.9	1.6
26	40	4.1	3.2	2.9	2.6	2.5	2.3	2.0	1.8	1.5
27	60	4.0	3.2	2.8	2.5	2.4	2.3	1.9	1.7	1.4
28	120	5.3	3.1	2.7	2.5	2.3	2.2	1.8	1.6	1.3
29	*	3.8	3.0	2.6	2.4	2.2	2.1	1.8	1.5	1.0

Nomograms:

A nomogram is constructed for the mechanical properties of geopolymer using decoded values of X_1 and X_2. Here, isoproperty lines are drawn for two variables, i.e., decoded value of temperature and particle size are plotted for constant values of response, keeping the curing time parameters at the lower decoded value. The particle size (X_2) is taken to be zero, the plot of X_1 vs. X_2 yielded a straight line. The isoproperty curves are shown in Figure 4.1. Figure 4.1 displays graph with coded value (X3 at lower value -1). Figure 4.2 shows graph with coded value (X2 at lower value 0). Figure 4.3 indicates graph with natural value (lower value, X3 = 20). Figure 4.4 presents graph with natural value (at base level X2 = 195).

The iso-strength lines shift to the right of the origin as indicated by the arrow in order of increasing value.

The nomogram has been drawn within the range of variation of variables that has been used in the present investigation. Since the equations are only valid in the range of composition selected, it was preferred not to use this equation for constructing nomograms beyond the range.

Morphologies of as-received pond ash and cured pond ash-based geopolymers are shown in Figures 4.5a and 4.5b. An FESEM image of as-received pond ash sample is shown in Figure 4.5a. Loosely packed elliptical particles with different aspect ratios are observed in the microstructure. Some porosity may also be observed in the particles [10]; whereas prepared pond ash-based geopolymer (Figure 4.5b) shows compacted mass. Compacted mass is formed due to curing of geopolymer material at 70 °C for different lengths of curing time. EDX analyses of both materials, i.e., as-received pond ash and prepared pond ash-based geopolymer, are shown in Figure 4.7.

The EDS image in Figure 4.6 (left) shows the main elements observed in as-received pond ash (Si, Al, Fe, Na, Mg, Ca and O); and those of GP prepared from pond ash are depicted in Figure 4.6 (right). The presence of all those elements shown in pond ash are also present in GP material formed from pond ash, with the only difference being their peak height.

Differential scanning calorimetry (DSC) is used to measure a number of characteristic properties of geopolymer pastes. It is likely to observe exothermic and endothermic events as well as glass transition temperatures (T_g). The DSC thermogram in Figure 4.7 shows the isothermal measurements of as-received PA and optimized GP product carried out in

Graph 1

Sl. no.	Temp (X_1)	Mesh size (X_2)	Time (X_3)	Y=16
1	0.1	0.34	-1	
2	0.2	0.022	-1	
3	0.3	-0.304	-1	
4	0.4	-0.63	-1	
				y=18
1	0.5	0.89	-1	
2	0.6	0.56	-1	
3	0.7	0.23	-1	
4	0.8	-0.08	-1	

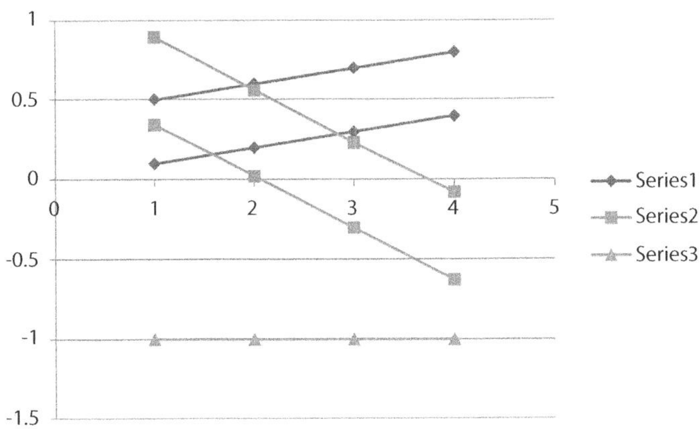

Figure 4.1 Graph with coded value (X3 at lower value -1).

Graph 2

Sl. no.	Temp (X_1)	Mesh size (X_2)	Curing time (X_3)	
1	0.3	0	-0.14	
2	0.4	0	0.33	Y=16
3	0.5	0	0.61	
4	0.6	0	0.8	
				y=18
1	0.8	0	-0.9	
2	0.85	0	-0.75	
3	0.9	0	-0.61	
4	0.95	0	-0.48	

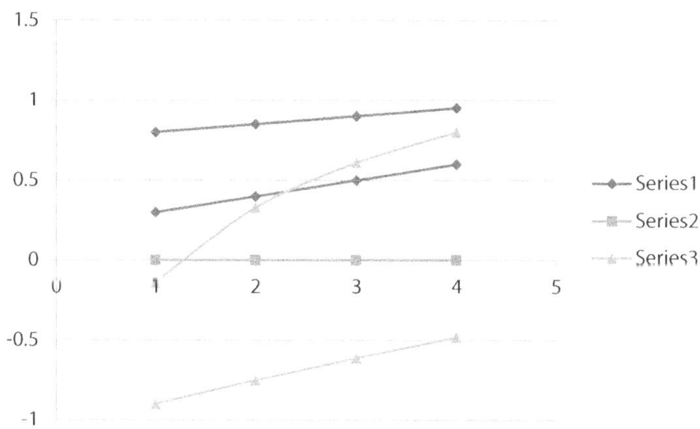

Figure 4.2 Graph with coded value (X_2 at lower value 0).

Graph 3

Sl. no.	Temp (X_1)	Mesh size (X_2)	Curing time (X_3)	
1	72	210.3	20	
2	74	195.99	20	Y=16
3	76	181.32	20	
4	78	166.65	20	
				y=18
1	80	235.05	20	
2	82	220.2	20	
3	84	205.35	20	
4	86	191.4	20	

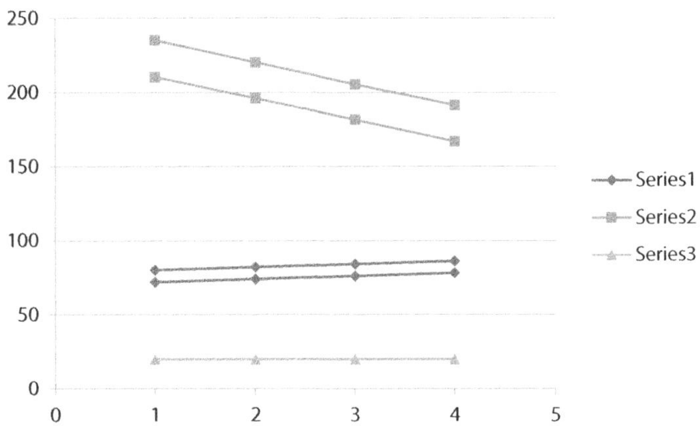

Figure 4.3 Graph with natural value (lower value, X_3=20).

Graph 4

Sl. no.	Temp (X_1)	Mesh size (X_2)	Curing time (X_3)	
1	76	195	23.44	
2	78	195	25.32	Y=16
3	80	195	26.44	
4	82	195	27.2	
				y=18
1	86	195	20.4	
2	87	195	21	
3	88	195	21.56	
4	89	195	22.08	

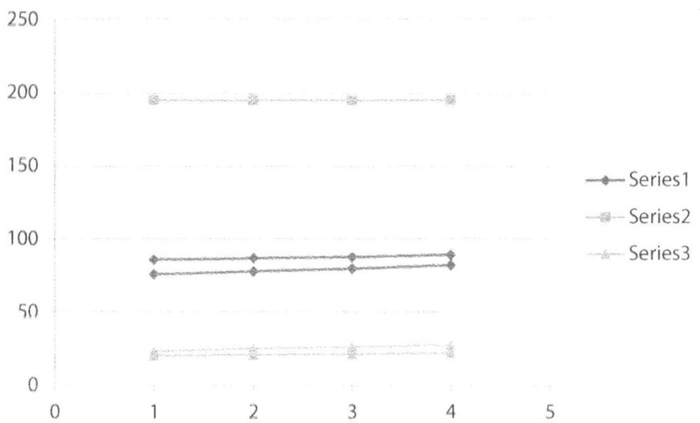

Figure 4.4 Graph with natural value (at base level X_2 = 195).

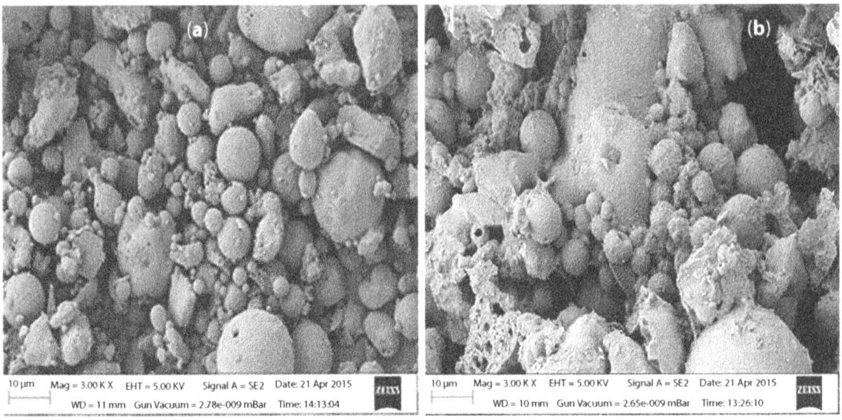

Figure 4.5 FESEM images of (a) As-received pond ash and (b) PA-based GP cured at 70 °C for 24 h with 240 meshes.

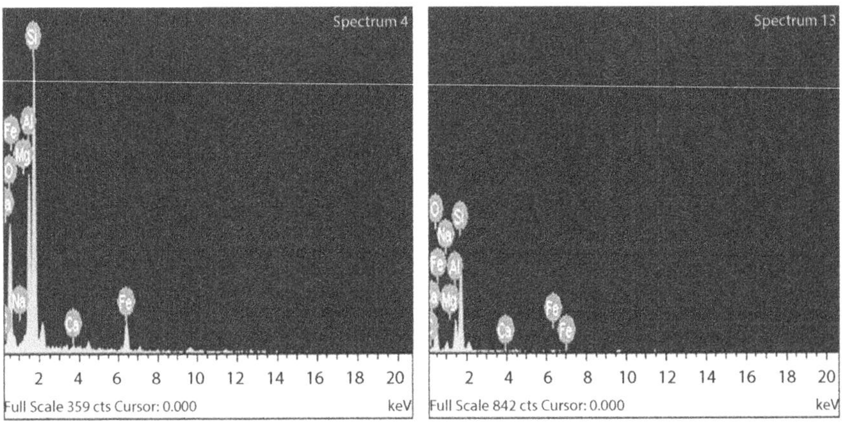

Figure 4.6 EDX images of elements in (left) As-received pond ash and (right) PA-based GP cured at 70 °C for 24 h.

the temperature range of 50–250 °C. In the thermogram, the as-prepared geopolymer shows exothermic peaks. A similar observation was made by Chindaprasirt and Rattanasak [11]. The DSC curve in Figure 4.7a indicates that the pond ash is smooth and there is no sign of geopolymeric reaction. However, the curve (b) of pond ash-based geopolymeric product demonstrates the occurrence of an exothermic peak at a temperature ~ 110 °C, which indicates that the geopolymeric reaction happened [12].

Figure 4.8 shows the weight loss of GP cured samples due to heating from 50–800 °C. Experiments were carried out under inert environment

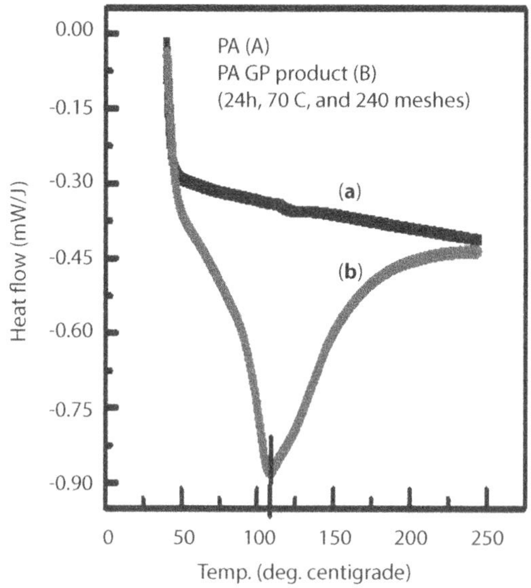

Figure 4.7 DSC isotherm curves of (a) as-received PA and (b) optimized GP product.

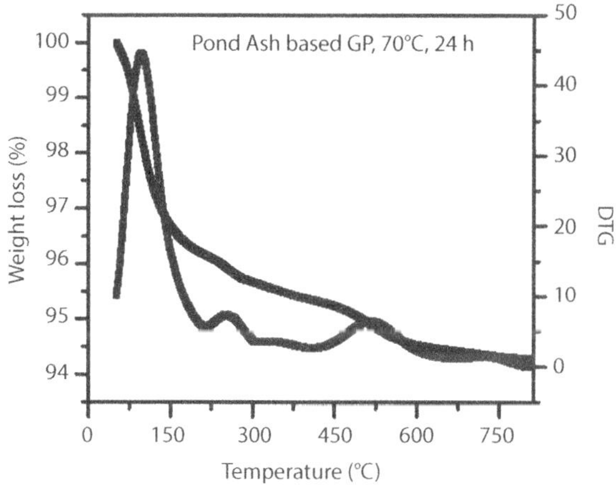

Figure 4.8 TGA plot of optimized GP product.

(N_2). It is observed that there is a sharp drop in the curve between 50 °C to 170 °C. This weight loss (3%) is due to removal of adsorbed moisture/water molecules [13]. A significant observation is the loss in water beginning around 150 °C till 450 °C temperature (humps in the thermogram), indicating loss of water which is present in the cage (formed between the polymer chains), since polymerization is completed during curing [14–17].

4.4 Conclusions

Statistical design of experiments is a satisfactory method for quantifying the effects of variables on mechanical property. An equation was obtained by applying a statistical design of the experiment. Significant coefficients of regression equation were eliminated by applying student's t-test. Adequacy regression was checked by 'F' test. Significant interactions probably provided a quantitative measure of the mechanical properties of geopolymer in a range of variations of variables. Regression equations were slightly nonlinear in nature (i.e., presence of significant binary and ternary coefficients in equations) but could satisfactorily predict the properties of variables within the narrow range of composition studied. A nomogram was constructed which helps the material designer select a suitable combination of variables to get desired properties. Optimization of mechanical properties of the geopolymeric process of pond ash was carried out with the addition of sodium silicate, sodium hydroxide and water-soluble plasticizer by adopting different process parameters, i.e., curing temperature, curing time and particle size. Optimum condition of preparation of geopolymeric product was 240 meshes, 70 °C curing temperature with 24 h curing through single factor experiment. This can form the base level of treatment combination around which the design matrix can be constructed. Optimum strength of developed in geopolymeric product was found to be 19 MPa by single factor experiment.

A DSC isotherm showed that the water of crystallization is found to be ~115 °C. The FESEM image of as-received pond ash and optimized geopolymeric product show morphological transformation from irregular spherical shape to compacted structure. EDX data indicate the presence of elements such as Si, Ca, Al, Fe, Na, Mg, and O.

From the results, it is concluded that for construction purpose, geopolymer can be suitable construction material to replace cement in cement-based products (such as brick and concrete). The material will be more sustainable in terms of environmental performance, economic viability and potential compared to OPC-based products.

Acknowledgments

The authors convey their sincere thanks to the Ministry of Mines, Government of India, for providing a grant to carry out the work [Grant number=F.No.:14/54/20214-Met.-IV dated: 29.12.2014]. They also convey their sincere thanks to GIET, Gunupur, Rayagada, Odisha, India, for providing lab facilities to do the research work. The authors would also like to thank the CRF, IIT Kharagpur, for providing testing facilities.

References

1. Production of bricks from waste materials – A review
2. M. Barbuta, R. Dana Bucur, S.M. Cimpeanu, G. Paraschiv, and D. Bucur, (2014), Wastes in Building Materials Industry, Vol. xxx, pp. xxx-xxx (DOI: 10.5772/59933).
3. H. Mehdizadeh and E.N. Kani, (**2018**), Modeling the Influence of Chemical Composition on Compressive Strength Behavior of Alkali-activated Phosphorus Slag Cement using Statistical Design, Canadian Journal of Civil Engineering, Vol. 45, pp. 1073-1083.
4. M.F. Triola, Elementary Statistics. 10 ed. **2007**: Addison Wesley.
5. Nazari, A., Riahi, S., and Bagheri, A., (**2012**), Designing Water Resistant Light Weight Geopolymers Produced from Waste Materials, Materials and Design, Vol. 35, pp. 296-302.
6. S. Riahi, A. Nazari, D. Zaarei, G. Khalaj, H. Bohlooli, and M.M. Kaykha, (**2012**), Compressive Strength of Ash-based Geopolymers at Early Ages Designed by Taguchi Method, Materials and Design, Vol. 37, pp. 443-449.
7. A. Nazari, H. Khanmohammadi, M. Amini, H. Hajiallahyari, and A. Rahimi, (**2012**), Production Geopolymers by Portland Cement: Designing the Main Parameters Effects on Compressive Strength by Taguchi Method, Materials and Design, Vol. 41, pp. 43-49.
8. M. Panigrahi, S. Mohanty, R.I. Ganguly1, R.R. Dash, (**2018**), An advanced cured high carbon ferrochrome slag (HCFCS) geopolymer (GP): A constructional materials, IOP Conf. Series: Materials Science and Engineering, Vol. 410, pp. 1-16 (DOI:10.1088/1757-899X/410/1/012002).
9. M. Panigrahi, S. Mohanty, R.R. Dash, R.I. Ganguly, (**2018**), Development of Novel Constructional Material from Industrial Solid Waste as Geopolymer, IOP Conf. Series: Materials Science and Engineering, Vol. 410, pp.1-12 (DOI:10.1088/1757-899X/410/1/012012).
10. R. Datta, S. K. Pal, (**2013**), Investigation on Bearing Capacity Behavior of Pond Ash Reinforced with Geotextile Grid, Electronic Journal of Geotechnical Engineering, Vol. 18, pp. 1631-1640.
11. P. Chindaprasirt and U. Rattanasak, (**2020**), Eco-production of Silica from Sugarcane Bagasse Ash for Use as A Photochromic Pigment Filler, Scientific Reports, Vol. 10: 9890, pp.1-8 (DOI: https://doi.org/10.1038/s41598-020-66885-y 1).
12. A.M. Al Mustafa Bakri, H. Kamarudin, M. Bnhussain, I.N. Khairul, A.R. Rafiza, and Y. Zarina, (**2012**), The Processing, Characterization, and

Properties of Fly Ash based Geopolymer Concrete, Reviews on Advanced Materials Science, Vol. 30, pp. 90-9.
13. F, Škvára L. Kopecky´ V. Šmilauer, and Z. Bittnar, (**2009**), Material and structural characterization of alkali activated low-calcium brown coal fly ash. J Hazard Mater, Vol. 168, pp. 711-20.
14. P. Chindaprasirt, S. Thaiwitcharoen, S. Kaewpirom, U. Rattanasak, (**2013**), Controlling Ettringite Formation in FBC Fly Ash Geopolymer Concrete, Cem Concr Compos, Vol. 41, pp. 24-8.
15. P.N. Lemougna, K.J.D. MacKenzie, and U.F.C. Melo, (**2011**), Synthesis and Thermal Properties of Inorganic Polymers (Geopolymers) for Structural and Refractory Applications from Volcanic Ash, Ceram Int, Vol. 37, pp. 3011-8.
16. F. Winnefeld A. Leemann M, Lucuk P. Svoboda and M. Neuroth, (**2010**), Assessment of Phase Formation in Alkali Activated Low and High Calcium Fly Ashes in Building Materials, Constr Build Mater, Vol. 24, pp. 1086-93.
17. P. Chindaprasirt and U. Rattanasak, (**2010**), Utilization of blended fluidized bed combustion (FBC) ash and pulverized coal combustion (PCC) fly ash in geopolymer. Waste Manage, Vol. 30, pp. 667-72.

5

Development of Pond Ash (PA)–High Carbon Ferrochrome (HCFC) Slag-Based Geopolymer Cementitious Materials

Muktikanta Panigrahi[1]*, Ratan Indu Ganguly[2] and Radha Raman Dash[3]

[1]PG Department of Materials Science, Maharaja Sriram Chandra Bhanja Deo University, Keonjhar Campus, Odisha, India
[2]Department of Metallurgical Engineering, National Institute of Technology, Raurkela, Odisha, India
[3]CSIR-National Metallurgical Laboratory, Jamshedpur, Jharkhand, India

Abstract

Pond ash (PA) is a reject of thermal power stations which contains silica (62.8%) and alumina (28.3%). However, these materials possess low-calcium. High-carbon ferrochrome (HCFC) slag has comparatively low silica (28%) as compared to the silica content of pond ash. Judicial mixing of these two materials is thought to be an important proposition for balancing alkali materials which help geopolymer (GP) formation have with good strength. In this century, researchers have put their efforts into geopolymer concrete. In this investigation, pond ash, high-carbon ferrochrome (HCFC) slag, sodium silicate and other ingredients, such as sodium hydroxide (NaOH) and water-soluble plasticizer (Sika), have been used to prepare novel GP products for use in heavy construction application(s). The mechanism involved in such polymer is that the silicon and the aluminum in the mixture of PA and HCFC slag react with an alkaline solution of sodium silicate, sand and aggregates with water-soluble plasticizer to form the geopolymer concrete that binds the non-reactive materials. Scanning electron microscope (SEM) analyses indicate that the mixture of crystalline and non-crystalline phases in HCFC slag may have beneficial effects when added with PA. In fact, enhanced mechanical properties, such as compressive strength, have been observed with the addition of slag to PA. DSC isotherm studies have indicated the formation of geopolymer; and TGA analysis has indicated the stability of prepared GP products. The structural

**Corresponding author*: muktikanta2@gmail.com

Muktikanta Panigrahi, Ratan Indu Ganguly and Radha Raman Dash (eds.) Development of Geopolymer from Pond Ash-Thermal Power Plant Waste: Novel Constructional Materials for Civil Engineers, (151–168) © 2023 Scrivener Publishing LLC

investigation and stoichiometric calculation were made by analytical testing like XRD and EDX. These observations suggest new-generation materials for construction engineers in the near future.

***Keywords*:** Pond ash, slag, geopolymer, strength properties, TGA

5.1 Introduction

Ordinary Portland cement (OPC) is a well-established material for construction purposes. However, from the perspective of pollution and ease of production, new materials are being considered to replace OPC. In the present scenario, in which the world economy is seriously being affected by environmental pollution, it is imperative that the new materials are developed that will dispense with the production of materials using energy-intensive processes [1–5]. Reports are available on global CO_2 emissions (at least 8%) during the production of OPC. Use of high kiln temperatures (i.e., 1450–1550 °C) during OPC production makes the process highly energy-intensive with the additional requirement of raw materials such as limestone. These materials are very essential for iron and steel industries [6–10].

With the above in mind, research has begun in the development of new-generation materials which can replace conventional materials such as cement. One such material is thought to be geopolymer (GP), which was developed and named by Professor Joseph Davidovits in the 1970s. Normally, geopolymer is synthesized using an alkali solution and an aluminosilicate powder. The reaction results in an alkali-aluminosilicate gel. Matched with OPC, geopolymer enjoys quick attainment of compressive strength [7, 11], lower permeability [12, 13], lower shrinkage [14, 15], and good resistance to acid and fire attack [16, 17]. Besides, geopolymer can be synthesized at a lower temperature if compared with the production temperature of OPC. Geopolymers are produced by the polymerization reaction occurring at room temperature [18]. Furthermore, almost no hazardous gases, such as SO*x*, NO*x*, or CO, are generated in the polymerization process [10]. Considering these merits, the scientific community has shown great interest in pursuing research for the development of geopolymer worldwide. Presently, geopolymers are used in different applications, *viz.*, novel building materials [19, 20], soil stabilization [21–23], immobilization of heavy metals [24–26], novel catalyst [27, 28], and adsorbent [29, 30].

It is prudent to analyze raw materials scientifically before proceeding to develop a process for formation of GP. For this purpose, XRD and SEM image analyses of raw materials are considered important.

Pond ash is a reject of thermal power stations which contains silica (62.8%) and alumina (28.3%) [31–40]. However, these materials possess low calcium. High-carbon ferrochrome (HCFC) slag has comparatively low silica (28%) as compared to silica content of pond ash [31–40]. Judicial mixing of these two materials is thought to be an important proposition for balancing alkali materials which help GP formation with good strength.

Some related reports are available where different waste raw materials are used for preparation of geopolymer; Chi and Huang [41] investigated the binding mechanism of alkali-activated FA/BFS with different ratios of FA to BFS. Compressive strength tests and flexural strength test were performed to evaluate the properties. They characterized structure with SEM and XRD. Their results showed that a fly ash/slag ratio has significant influence on binding mechanism.

Abdelkader *et al.* [42] used FA, BFS, and sodium carbonate (Na_2CO_3) to prepare blends. For preparing the blend, Na_2CO_3 was used as an alkali activator. The strength property of geopolymer depends on the proportion of FA/BFS blends. They observed that mixture containing 25% FA and 75% BFS yielded optimum properties after 90 days of curing. Increasing FA content to 50% led to a significant drop in strength.

Phoo-ngernkham *et al.* [43] examined the effects of alkaline solutions, i.e., NaOH and Na_2SiO_3 solutions, on the performance of FA/BFS geopolymer. Results showed good properties of geopolymer. It was concluded that the properties of geopolymer depend on the appropriate proportion of source materials treated with alkali activators. The compressive strength of geopolymer was found to have increased due to the formation of additional calcium silicate hydrate (CSH).

Sankar *et al.* [44] prepared Na_2SiO_3-activated slag-fly ash-based geopolymers and explored their performance such as compressive and flexural strengths. They found that increasing the slag/fly ash ratio resulted in an enhancement of strength properties due to the formation of calcium silicate hydrate (CSH), calcium aluminum silicate hydrate (CASH), and Ca- and Na-based geopolymers.

Yu *et al.* [45] prepared alkali-activated-based geopolymer under normal temperature or at higher temperature conditions. Since the slag-based geopolymer reacts quite rapidly at normal temperature, the depolymerization and polycondensation processes can be properly separated at low temperature conditions, which was beneficial for the study of the reaction process and mechanism of the alkali-based geopolymers. They evaluated the performance of the alkali-based geopolymer under vacuum and freeze-thaw cycles conditions. Also, they advocated recycling and reusing wastewater.

Qiu *et al.* [46] investigated the potential of fly ash (FA)/blast furnace slag (BFS)-based geopolymer as a novel backfilling material. They explored the effects of NaOH concentration and FA/BFS mass ratios. Other physical properties were characterized using XRD, FTIR, and TG-DTG analyses. Their results indicated that the reaction products and strengths of geopolymer depend on NaOH concentration and the source of raw materials used for its preparation. Slump, final setting time, and setting ratio were increased in the mixture of FA-based geopolymer. However, increases in FA content reduced compressive strength. Microstructure of backfilling material (BM) indicated lower reactivity compared to BFS. Microstructure analyses revealed that the matrix tends to be denser with BFS content and increased NaOH concentration.

Mallika *et al.* [47] prepared geopolymer for replacing cement material using aluminum and silicon-bearing materials and activated alkaline solution. They also evaluated workability, mix proportions (with different binder materials), mechanical performance (compressive and tensile strength) and durable properties (acid attack, high-temperature resistance, shrinkage resistance). The results showed that geopolymer is an environmentally friendly and feasible material for structures, which may be adopted as a suitable material to replace conventional cement in the near future.

Kurtoğlu *et al.* [48] prepared both fly ash-based geopolymer concretes (FAGPC) and slag-based geopolymer concretes (SGPC) using alkaline solution, i.e., mixture of sodium silicate (Na_2SiO_3) and sodium hydroxide (NaOH) solutions. Ordinary Portland cement (OPC)-based concrete was also prepared for comparison purpose. They investigated mechanical and short-term durability properties of both geopolymer and cement-based concretes. These studies aimed to make a contribution to the standardization process of the geopolymer concretes useful in the construction industry. They exposed SGPC, FAGPC and OPC concretes to sulfuric acid (H_2SO_4), magnesium sulfate ($MgSO_4$) and sea water (NaCl) solutions with concentrations of 5%, 5% and 3.5%, respectively, for studying corrosion behavior. The results of SGPC (100% slag) showed better properties compared to the other two concretes.

Ding *et al.* [49] prepared geopolymer paste as a repair material for existing concrete structure. They used ground granulated blast furnace slag and coal fly ash for preparation of the geopolymer. Properties such as compressive strength, flexural strength, and bonding strength between the slag/fly ashes-based geopolymer paste were evaluated. The nature of the geopolymeric reaction (i.e., bonding nature) was examined using FTIR and NMR analyses. Compressive strength and flexural strength of the geopolymer paste were found to be 47 MPa and 16 MPa, respectively. It was concluded that slag/fly ash-based geopolymer paste has potential for technological applications in the near future.

Sindhunata *et al.* [50] synthesized geopolymers using class F fly ash. They studied structure and pore distribution in materials using an electron microscope and porosimeter. The Si/Al ratio played an important role in controlling porosity of the structure. It was varied from1.51 to 2.24. They also varied curing temperature and the silicate ratio in activating solutions (SiO/MO, M = Na or K). The result(s) indicated that pore structure of K-based geopolymer is more susceptible to change in temperature than that of Na-based geopolymer. A higher Si/Al ratio and finer pores were obtained in geopolymers synthesized at higher temperature and silicate ratios. Elevating curing temperature increased the extent and rate of reaction as shown through an increase in mesoporous volume, surface area, and an accelerated setting time. The kinetics appeared to be temperature-controlled only before the material hardened. Very high silicate ratios (SiO/MO $\geq$ 2.0) were also believed to have slowed the reactions. The pore structure of K-based geopolymer was more susceptible to change with temperature than that of Na-based geopolymer.

Xie *et al.* [51] developed ultra-high-performance concrete from coarse aggregates, ground granulated blast furnace slag (GGBS) and fly ash using modern green engineering technologies. The developed green concrete indicated improved sulfate resistance. Different characterization techniques were used to determine the physical and mechanical properties of the concrete, including compression testing for strength, and X-ray diffraction (XRD) and scanning electron microscopy (SEM) for structure. Different exposure cycles in a sulfate environment were also performed. The residual compressive strength of the concrete was shown to be higher after sulfate exposure.

Karakoc *et al.* [52] produced geopolymer cements using ferrochrome slag (FS) through alkali activation (NaOH and Na_2SiO_3) process. They studied alkali dosage effects (silica modulus when using sodium metasilicate solution), setting time, hydration heat and compressive strength. The setting time varied between 120 and 870 min, which showed variability depending on Na_2O content. The highest 28-day compressive strength of the geopolymer paste was obtained from 7% Na_2O concentration and 0.70 silica modulus. Geopolymer mortars were prepared using combination (FS: sand: alkali activator ratio=1:2:0.30) for determining the compressive strength and were cured at 60 °C and 80 °C kept for 20 h. The highest 28-day strength of the geopolymer mortar was obtained at 0.30 w/b ratio. The heat of hydration of geopolymer paste samples was found to be less than normal Portland cements. Scanning electron microscopy (SEM) and X-ray diffraction (XRD) were employed to study the microstructural properties of the geopolymers.

Jena and Panigrahi have [53] prepared ferrochrome slag-based geopolymer concrete and controlled geopolymer concrete. They evaluated the performance, structure (XRD and FTIR), and microstructure (SEM with EDX) of the geopolymer concrete.

Falayi [54] synthesized geopolymer using ferrochrome (FeCr) slag as a precursor with alkaline solution. They studied the effect of KOH concentration, liquid solid ratio (L/S), content of potassium metasilicate (KS) or potassium aluminate (KA), curing time on compressive strength and metal leachability on synthesized geopolymer. A 10 M KOH and L/S of 0.26 yielded a geopolymer with a compressive strength 13.0 MPa after curing for 28 days at ambient temperature. KS: KOH (0.125 wt) yielded geopolymer with a compressive strength 14.7 MPa whilst 1.25 wt KA: KOH addition yielded a geopolymer with compressive strength 24.5 MPa. The increase in strength was due to the formation of calcium silicate hydrate. The activated FeCr slag geopolymer was found to be the most competent among all other geopolymers.

In the present work, high-carbon ferrochrome slags (HCFC slag) and pond ash (PA) were used as raw materials. They were activated by sodium hydroxide (NaOH) solution having different concentrations during the preparation of geopolymer. Combined effects of PA and HCFC slag treated with activated alkaline solution on the geopolymer properties were studied. Different phases were identified by X-ray diffraction pattern. FTIR spectra were used to detect different absorption bands and a TG-DTG study to explore the stability. DSC analysis was used in the experiment to discover the condensation polymerization; and surface topography was measured by SEM.

5.2 Experimental Details

5.2.1 Source of Materials

Pond ash (PA) was collected from NALCO Navaratna Company, Damanjodi, Odisha India. High-carbon ferrochrome (HCFC) slag, the main ingredient of geopolymer, was collected from Shyam Ferro Alloys (Durgapur, West Bengal, India). Sodium hydroxide (NaOH) was used as an activator, which was obtained from M/S Loba Chemicals, India. Sodium silicate (Na_2SiO_3) was used for the purpose of polymerization and was procured from Merck,

India. Water-soluble plasticizer (Sika) was purchased from a reputed firm located in the southern part of India (Visakhapatnam market). Plasticizer usually enhances plasticity of the product, i.e., geopolymer.

5.2.2 PA/HCFC Slag-Based Geopolymer (GP) Preparation

Steps in the geopolymer preparation process were previously described in Chapter 3. In the preparation of PA/HCFC slag materials, both pond ash and high-carbon ferrochrome (HCFC) slag were used in the mixture [55, 56]. Steps of PA/HCFC slag-based GP preparation are described below [55, 56]:

Step 1: Raw Material Preparation
Geopolymer is prepared from pond ash, HCFC slag, sodium silicate, and alkali. HCFC slag is crushed and ground to 63 micron size. Fine particles are then dried in an oven for 2 h at 120 °C. Sieved sand is used for preparing both geopolymer mortar and geopolymer concrete. Geopolymer concrete is prepared by mixing PA/HCFC-based GP, sand, and aggregates.
Step 2: Geopolymer Preparation with Raw Materials
The ground and dried materials are mixed with requisite amount of NaOH (8 M), Na_2SiO_3 and water-soluble plasticizer (Sika) (Table 5.1).
Step 3: Molding, Casting and Compaction
The semi-solid masses are placed in a mold which is cylindrical in shape (55 mm diameter and 70 mm height). The mixture is rammed in the mold, the uniformity of which is ensured by continuously ramming the mass in the mold three times. Following ramming of the mass in the mold, the samples are left for 3 min for setting before demolding the green sample.
Step 4: Curing Process
The green samples are cured at different temperatures, i.e., 60, 70, 80 °C for different lengths of time, i.e., 7 days and 28 days.

Table 5.1 PA/HCFC slag-based GP (three samples with same compositions).

Sample code	PA (51%) + HCFC slag (34%)	Sod. silicate (SS)	Alkali (8 M, NaOH) (SH)	Water-soluble plasticizer (Sika)
S1	85%	12%	3%	1–2 ml
S2	85%	12%	3%	1–2 ml
S3	85%	12%	3%	1–2 ml

5.2.3 PA/HCFC-Based Geopolymeric Mortar and Concrete

For making GP mortar or GP concrete, materials were mixed with sand/aggregates in suitable proportion and molded in a die. The geopolymer samples were then cured at 70 °C for 24 h. The compositions of geopolymer mortar (geopolymer mixture and sand) and concrete (geopolymer mixture, sand and aggregates) are presented in Table 5.2 and Table 5.3, respectively.

5.2.4 Characterizations of PA/HCFC-Based Geopolymeric Material

In this study, compressive strength test was performed on 70 × 50 mm cylindrical specimens using a 20 KN digital compressive testing machine (Mechatronics Control System; Model: MECH-CS/UTE 20T). A set of three samples, described in Table 5.3, was chosen from one lot. The experiments were replicated in batches to ascertain the validity of test results.

Use of scanning electron microscopy (SEM) and energy-dispersive X-ray (EDX) analysis helped to understand the morphology and chemical compositions of the products.

The thermal stability of the PA/HCFC slag-based GP and its GP mortar/concrete cured at 70 °C for 24 h was determined by thermogravimetric analysis.

The enthalpy of fusion for as-received PA, HCFC slag, PA/HCFC slag-based GP and its GP mortar/concrete cured at 70 °C for 24 h was estimated using differential scanning calorimetry (DSC). The parameters, such as heating rate, temperature range and nitrogen environment (N2), were set as 10 °C/min and 50–300 °C, respectively.

Table 5.2 PA/HCFC slag-based GP mortar (three samples with same compositions (Mortar=1.0:3.0).

Sample code	PA (12.675%)/ HCFC slag (8.45%) + sand (63.375%)	Sod. silicate (SS)	Alkali (8 M, NaOH) (SH)	Water soluble plasticizer (Sika)
S1	85%	12%	3%	1–2 ml
S2	85%	12%	3%	1–2 ml
S3	85%	12%	3%	1–2 ml

Table 5.3 PA/HCFC slag-based GP concrete (three samples with same compositions (Concrete=1.0:1.5:3.0).

Sample code	PA (9.218184%)/ HCFC slag (6.145456%) + sand (23.04546%) + aggregates (46.09092)	Sod. silicate (SS)	Alkali (8 M, NaOH) (SH)	Water-soluble plasticizer (Sika)
S1	85%	12%	3%	1–2 ml
S2	85%	12%	3%	1–2 ml
S3	85%	12%	3%	1–2 ml

Prepared samples were then characterized by XRD, FTIR, TGA, DSC, etc.

5.2.5 Results and Discussion

Table 5.4 shows the compressive strength of PA/HCFC Slag GP materials prepared with different compositions. These materials were prepared using optimal treatment combination parameters. The average strength level was 22 MPa for PA/HCFC slag-based GP. This is equivalent to standard M20 grade concrete used for making structures [55, 56].

Results and discussion on strength properties for getting two different new geopolymer materials, namely PA/HCFC Slag GP materials (mortar and concrete) were evaluated. PA/HCFC Slag GP mortar and PA/HCFC Slag GP concrete showed 26.4 MPa and 34 MPa, respectively. Both the materials used an optimized combination of process parameters such as composition, molar concentration of alkalis, curing time and curing temperature [55, 56].

It is essential to understand the process of strengthening and characterizing these materials (GP slag, GP slag mortar, GP slag concrete).

Figures 5.1a and 5.1b depict SEM micrographs with EDX analysis of PA/HCFC-based GP. The micrograph of PA/HCFC slag-based GP cured at 70 °C for 24 h is shown in Figure 5.1a. It has enabled denser agglomeration of powder. EDX analysis (Figure 5.1b) confirms the presence of elements, such as aluminium, silicon, and oxygen, which support the formation of geopolymer with Si, Al, Fe, Mg, Ca, Cr, Na, Ti, O, etc.

Table 5.4 Compressive strength of PA/HCFC slag GP materials.

S. no.	Sample type with combinations	Optimal parameters	Av compressive strength (MPa)
1	PA/HCFC Slag GP (PA+HCFC slag = 85%)	24 h 70 °C oven curing. 12% 8M NaOH, 3% Na_2SiO_3, 1-2 mL Sika	22
2	PA/HCFC Slag GP Mortar (PA+HCFC slag + sand = 85%)	24 h 70 °C oven curing. 12% 8M NaOH, 3% Na_2SiO_3, 1-2 mL Sika	26.4
3	PA/HCFC Slag GP Concrete (PA+HCFC slag + sand + Aggregates = 85%)	24 h 70 °C oven curing 12% 8M NaOH, 3% Na_2SiO_3, 1-2 mL Sika	34

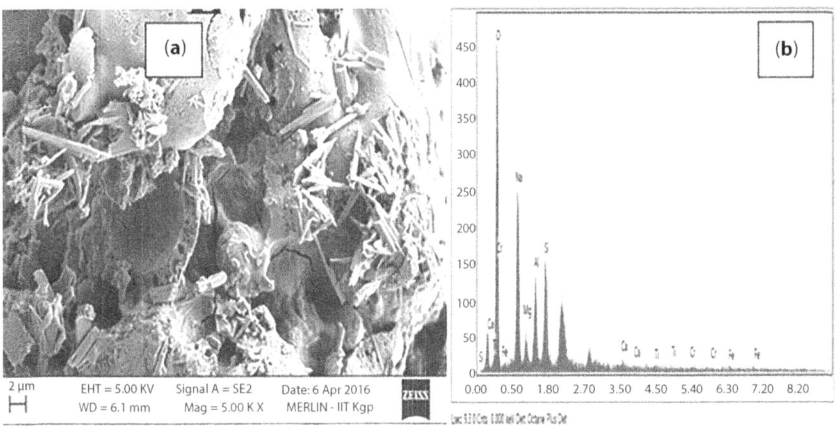

Figure 5.1 SEM image (in high resolution) of PA/HCFC based GP (a), and its EDS analysis (b).

A similar micrograph is observed for PA/HCFC slag-based GP mortar (Figure 5.2a). Needle-shaped phases are widespread on the surface of the particles. The needles have high aspect ratios. The EDX analysis confirmed the presence of elements, such as aluminium, silicon, and oxygen, which support formation of geopolymer with Si, Al, Fe, Mg, Ca, Cr, Na, Ti, O, etc.

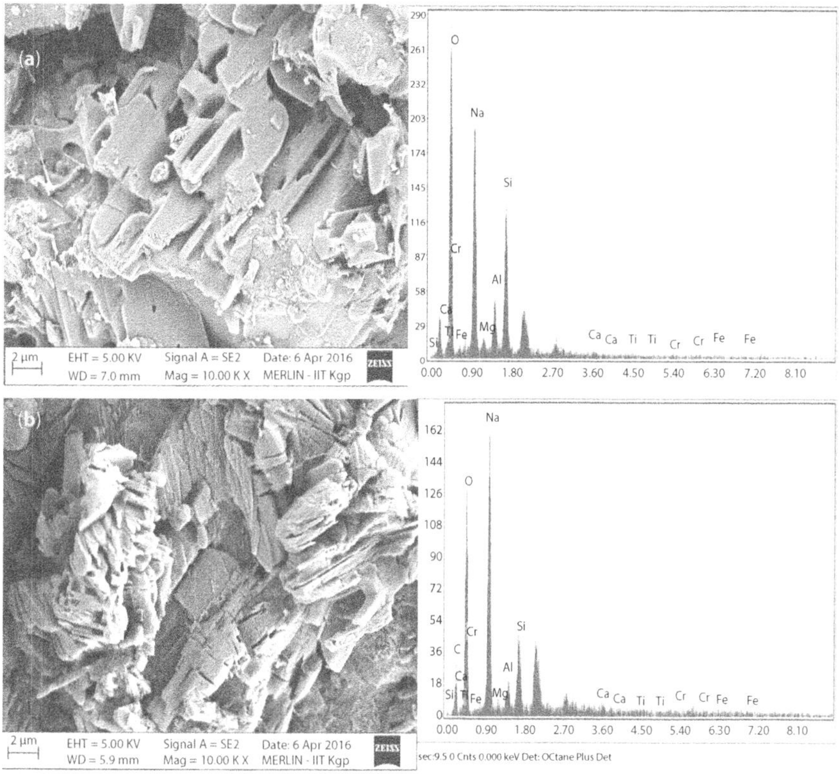

Figure 5.2 SEM image (in high resolution) of (a) PA/HCFC-based GP mortar and its EDX and (b) PA/HCFC-based GP concrete and its EDX.

The topography of PA/HCFC slag-based GP concrete (Figure 5.2b) shows the presence of compact rod-shaped phases widespread on the surface of the particles. The EDX analysis confirmed the presence of elements, such as aluminium, silicon, and oxygen, which support the formation of geopolymer with Si, Al, Fe, Mg, Ca, Cr, Na, Ti, O, etc.

Thus, because of structural difference, load-bearing capacities of the different composites show the different strength. It is presumed that the GP slag concrete materials showed the best mechanical properties because of different microstructural constituents. The load-bearing capacity of additives in PA/HCFC slag GP concrete have much better capacity to absorb load usually found in cement concrete. All the above results are encouraging for further work for replacing conventional cement.

Figure 5.3 (a & b) show two thermograms plotted for GPs (PA GP and HCFC slag GP). In both the plot of as-received pond ash as well as HCFC

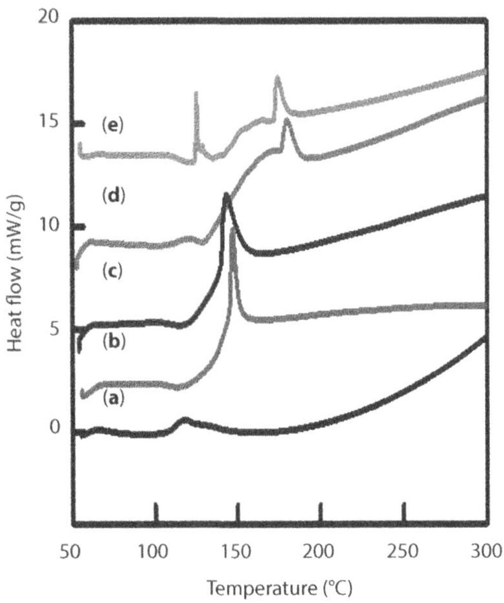

Figure 5.3 DSC isotherm of (a) PA-based GP, (b) HCFC-based GP, (c) PA/HCFC-based GP, (d) PA/HCFC-based GP motar, and (e) PA/HCFC-based Concrete GP.

slag are superimposed for comparison purpose. The figure clearly shows endo-thermic and exo-thermic peaks occurring at two different temperatures, i.e., 110 °C and 150 °C, respectively. This is due to oozing of water molecules attached to the cage of polymer chains formed during curing where condensation polymerization occurred [55, 56].

Thermograms were obtained for other varieties of GPs, i.e., PA/HCFC slag-based GP, PA/HCFC slag-based GP mortar, and PA/HCFC slag-based GP concrete (Figures 5.3a–e). Prominent isotherm peaks are observed for all the above materials at around 150 °C. Addition of PA to HCFC slag-based GP shifted endo-thermic peak to a lower temperature. However, there is only a slight shift of exo-thermic peak. This is presumably due to addition of PA in the later GP (HCFC slag-based GP).

The PA/HCFC slag-based GP mortar showes shift of peaks to a lower temperature. The occurrence of multiple peaks is believed to be due to different geopolymerization reactions occurring due to addition of sand.

A similar trend was observed for other composition prepared with aggregates, called concrete. Comparative value of loss in mass were recorded to be the least for PA/HCFC slag-based GP materials cured at 70 °C for 24 h (Figure 5.4c). The worst affected material was PA-based GP

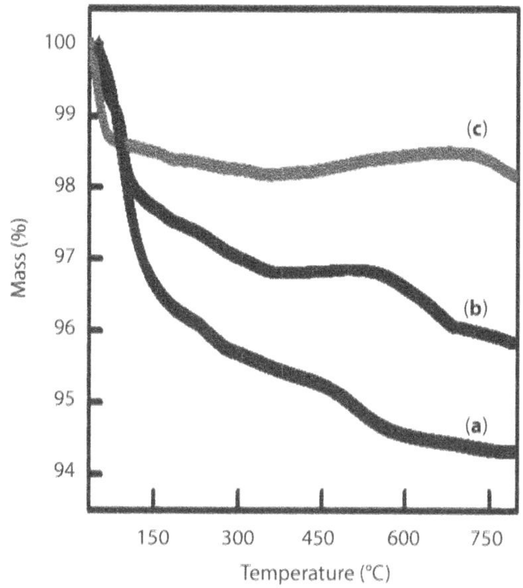

Figure 5.4 TGA curve of (a) PA GP cured at 70 °C for 24 h, (b) HCFC slag and (c) PA/HCFC slag-based GP cured at 70 °C for 24 h.

(Figure 5.4a). Relative stability was better for HCFC slag-based GP as compared to PA-based GP. This is attributed to the relative stability of structures built due to polymerization [55, 56].

5.3 Conclusions

Parameters for treatment combination were found to be sodium silicate (12%), sodium hydroxide (3%, 8M), and water-soluble plasticizer (1–2 mL), curing temperature (70 °C) and curing time (24 h), respectively. Maximum strength of PA/HCFC slag-based GP, PA/HCFC slag-based GP motor, and PA/HCFC slag-based GP concrete was observed to be 22 MPa, 26.4 MPa, 34 MPa, respectively. All three materials were characterized with SEM/EDX analysis. Some differences in microstructure were observed for three materials. PA/HCFC slag-based GP showed needle-shaped crystalline phase with occurrence of cleavage surface. PA/HCFC slag-based GP motor showed denser agglomeration, which indicated the presence of higher amounts of crystalline phases that occurred due to polymerization reaction. PA/HCFC slag-based GP concrete showed compact, widespread rod-shaped crystalline phases with unreacted particles. Crystalline phases were less prominent with appearance of glassy phase in the matrix. EDX analysis showed the presence of desired elements. Based on compressive strength data, it was evident that Pond Ash/HCFC slag-based GP, Pond Ash/HCFC Slag/Sand or Pond Ash/HCFC Slag/Sand/Aggregates was comparable to that of M25 grade concrete or fly ash-based geopolymer concrete. Well-supported thermal data was documented for making structural materials. In addition to their suitable mechanical properties, it is assumed that the building products, such as brick and concrete, will be more sustainable in terms of environmental performance, economic viability and potential for waste utilization.

Acknowledgments

The authors convey their sincere thanks to the Ministry of Mines, Government of India, for providing a grant to carry out the work [Grant number= F.No.:14/54/20214-Met.-IV dated: 29.12.2014]. The authors also convey their sincere thanks to GIET, Gunupur, Rayagada, Odisha, India, for providing lab facilities to do the research work. The authors would also like to thank CRF, IIT Kharagpur, for providing testing facilities.

References

1. P. Benito, C. Leonelli, V. Medri, and A. Vaccari, (2013), Geopolymers: a New And Smart Way for A Sustainable Development, Applied Clay Science, Vol. 73, p. 1-xxx.
2. H.A. Abdel Gawwad, S. Abd El-Aleem, and A.S. Ouda, (2016), Preparation and Characterization of One-Part Non-Portland Cement, Ceramics International, Vol. 42, pp. 220-228.
3. W. Zhu, X. Chen, A. Zhao, L. J. Struble, and E.-H. Yang, (2019), Synthesis of High Strength Binders from Alkali Activation of Glass Materials from Municipal Solid Waste Incineration Bottom Ash," Journal of Cleaner Production, Vol. 212, pp. 261-269.
4. L. Yang, E. Yilmaz, J. Li, H. Liu, and H. Jiang, (2018), Effect of Superplasticizer Type and Dosage on Fluidity and Strength Behavior of Cemented Tailings Backfill with Different Solid, journal name, and Vol.xxx, pp. 290-298.
5. J.-P. Qiu, L. Yang, J. Xing, and X.-G. Sun, (2018), Analytical Solution for Determining the Required Strength of Mine Backfill based on Its Damage Constitutive Model, Soil Mechanics and Foundation Engineering, Vol. 54, pp. 371-376.

6. F. Slaty, H. Khoury, J. Wastiels, and H. Rahier, (2013), Characterization of Alkali Activated Kaolinitic Clay, Applied Clay Science, Vol. 75-76, pp. 120-125.
7. I. Ozer and S. Soyer-Uzun, (2015), Relations Between the Structural Characteristics and Compressive Strength in Metakaolin Based Geopolymers with Different Molar Si/Al Ratios, Ceramics International, Vol. 41, pp. 10192-10198.
8. L. Yang, J. Qiu, H. Jiang, S. Hu, H. Li, and S. Li, (2017), Use of Cemented Super-Fine Unclassified Tailings Backfill for Control of Subsidence, Minerals, Vol. 7, p. 216-xxx.
9 J. Qiu, L. Yang, X. Sun, J. Xing, and S. Li, (2017), Strength Characteristics and Failure Mechanism of Cemented Super-Fine Unclassified Tailings Backfill, Minerals, Vol. 7, p. 58-xxx.
10. M. M. Yadollahi, A. Benli, and R. Demirboga, (2015), +e Effects of Silica Modulus and Aging on Compressive Strength of Pumice based Geopolymer Composites, Construction and Building Materials, Vol. 94, pp. 767-774.
11. H. Cheng, K.-L. Lin, R. Cui, C.-L. Hwang, T.-W. Cheng, and Y.-M. Chang, (2015), Effect of Solid-to-Liquid Ratios on the Properties of Waste Catalyst-Metakaolin based Geopolymers," Construction and Building Materials, Vol. 88, pp. 74-83.
12. I. Ismail, S.A. Bernal, J.L. Provis, (2013), Influence of Fly Ash on the Water and Chloride Permeability of Alkali-Activated Slag Mortars and Concretes, Construction and Building Materials, Vol. 48, pp. 1187-1201.
13. J. Wongpa, K. Kiattikomol, C. Jaturapitakkul, and P. Chindaprasirt, (2010), Compressive Strength, Modulus of Elasticity, and Water Permeability of Inorganic Polymer Concrete, Materials and Design, Vol. 31, pp. 4748-4754.
14. Y. Ma and G. Ye, (2015), +e Shrinkage of Alkali Activated Fly Ash, Cement and Concrete Research, Vol. 68, pp. 75-82.
15. N. K. Lee, J. G. Jang, and H.K. Lee, (2014), Shrinkage Characteristics of Alkali-Activated Fly Ash/Slag Paste and Mortar at Early Ages, Cement and Concrete Composites, Vol. 53, pp. 239-248.
16. S.A. Bernal, E.D. Rodríguez, R. Mejía de Gutiérrez, and J. L. Provis, "Performance of alkali-activated slag mortars exposed to acids," Journal of Sustainable Cement-Based Materials, vol. 1, no. 3, pp. 138–151, 2012.
17. R.R. Lloyd, J.L. Provis, and J.S.J. Van Deventer, (2011), Acid Resistance of Inorganic Polymer Binders. 1. Corrosion rate, Materials and Structures, Vol. 45, pp. 1–14.
18. K. Komnitsas and D. Zaharaki, (2007), Geopolymerization: A Review and Prospects for the Minerals Industry, Minerals Engineering, Vol. 20, pp. 1261-1277.
19. Y. Yuyou, C. Zengdi, L. Xiangqian, and D. Haijun, (2015), Development and Materials Characteristics of Fly Ash-Slag-Based Grout for Use in Sulfate-Rich Environments, Clean Technologies and Environmental Policy.

20. L.R. Steiner, A.M. Bernardin, and F. Pelisser, (2015), Effectiveness of Ceramic Tile Polishing Residues as Supplementary Cementitious Materials for Cement Mortars, Sustainable Materials and Technologies, Vol. 4, pp. 30-35.
21. Y. Yi, X. Zheng, S. Liu, and A. Al-Tabbaa, (2015), Comparison of Reactive Magnesia- and Carbide Slag-Activated Ground Granulated Blast Furnace Slag and Portland Cement for Stabilization of A Natural Soil, Applied Clay Science, Vol. 111, pp. 21-26.
22. G.N. Obuzor, J.M. Kinuthia, and R.B. Robinson, (2012), Soil Stabilization with Lime-Activated-GGBS-A Mitigation to Flooding Effects on Road Structural Layers/Embankments Constructed on Floodplains, Engineering Geology, Vol. 151, pp. 112–119.
23. P.P. Falciglia, S. Cannata, S. Romano, and F. G.A. Vagliasindi, (2014), Stabilisation/Solidification of Radionuclide Polluted Soils-Part I: Assessment of Setting Time, Mechanical Resistance, γ-Radiation Shielding and Leachate c-Radiation, Journal of Geochemical Exploration, Vol. 142, pp. 104–111.
24. L. Zheng, W. Wang, and Y. Shi, (2010), Effects of Alkaline Dosage and Si/Al Ratio on the Immobilization of Heavy Metals in Municipal Solid Waste Incineration Fly Ash-based Geopolymer, Chemosphere, Vol. 79, pp. 665-671.
25. D. Li, X. Guo, Z. Xu, Q. Tian, and Q. Feng, (2015), Leaching Behavior of Metals from Copper Anode Slime using an Alkali Fusion-Leaching Process, Hydrometallurgy, Vol. 157, pp. 9-12.
26. V. Nikoli, M. Komljenovi, N. Marjanovi, Z. Baščarevi, and R. Petrovi, (2014), Lead Immobilization by Geopolymers based on Mechanically Activated Fly Ash, Ceramics International, Vol. 40, pp. 8479–8488.
27. Y.J. Zhang and Q. Chai, (2014) Alkali-Activated Blast Furnace Slag based Nanomaterial as a Novel Catalyst for Synthesis of Hydrogen Fuel, Fuel, Vol. 115, pp. 84-87.
28. P. Sazama, O. Bortnovsky, J. Dedecek, Z. Tvaruzkov, and Z. Sobalık, (2011), Geopolymer based Catalysts-New Group of Catalytic Materials, Catalysis Today, Vol. 164, pp. 92-99.
29. M.S. Al-Harahsheh, K. Al Zboon, L. Al-Makhadmeh, M. Hararah, and M. Mahasneh, (2015), Fly Ash based Geopolymer for Heavy Metal Removal: A Case Study on Copper Removal, Journal of Environmental Chemical Engineering, Vol. 3, pp. 1669-1677.
30. M.N. Muzek, S. Svilovic, and J. Zelic, (2013), Fly Ash-based Geopolymeric Adsorbent for Copper Ion Removal from Waste Water, Desalination and Water Treatment, Vol. 52, pp. 2519–2526.
31. J. Davidovits, Geopolymer Chemistry and Applications (Geopolymer Institute, Saint-Quentin, France, 2008.
32. H. Xu and J.S.J. Van Deventer, (1999), The Geopolymerization of Natural Aluminosilicate, Proceedings: 2nd International Conference on Geopolymer, 99 (Geopolymer Institute, pp. 43-63.

33. A. Palomo M.W. Grutzeck, and M.T. Blanco-Varela, (1999), Alkali Activated Fly Ashes: Cement for the Future, Cement and Concrete Research, Vol. xxx, pp. 1323-1329.
34. H. Xu and J.S.J. Van Deventer, (2002), Geopolymerization of Multiple Minerals (Minerals Engineering, Vol. xxx, pp. 1131-1139.
35. A.M. Fernandez-Jimenez, E.E.A. Lachowski Palomo, and D.E. Macphee. (2004), Microstructural Characterization of Alkali-activated PFA Matrices for Waste Immobilization, Cement and Concrete Composites, Vol. xxx, pp. 1001-1006.
36. T. Bakharev, (2005), Geopolymeric Materials Prepared using Class F Fly Ash and Elevated Temperature Curing, Cement and Concrete Research, Vol. xxx, pp. 1224-1232.
37. S. Andini, R. Cioffi, F. Colangelo, T. Grieco, F. Montagnaro, and L. Santoro, (2008), Coal Fly Ash as Raw Material for the Manufacture of Geopolymer-based Products, Waste Management, Vol. xxx, pp. 416-423.
38. P. Chindaprasirt, T. Chareerat, and V. Sirivivatnanon, (**2007**), Workability and Strength of Coarse High Calcium Fly Ash Geopolymer, Cement and Concrete Composites, Vol. xxx, pp. 224-229.
39. A. Sathonsaowaphak, P. Chindaprasirt, and K. Pimraksa, (**2009**), Workability and Strength of Lignite Bottom Ash Geopolymer Mortar, Journal of Hazardous Materials, Vol. xxx, pp. 44-50.
40. J.G.S. Van-Jaarsveld, J.S.J. Van Deventer, and L. Lorenzen, (xxx), The Potential Use of Geopolymeric Materials to Immobilize Toxic Metals Part I, Theory and Applications, Mineral Engineering, Vol. 10, pp. 659-669.
41. M. Chi and R. Huang, (**2013**), Binding Mechanism and Properties of Alkali-Activated Fly Ash/Slag Mortars, Construction and Building Materials, Vol. 40, pp. 291-298.
42. A.F. Abdelkader, F. Jin, and A. Al-Tabbaa, (**2016**), Development of Greener Alkali-Activated Cement: Utilisation of Sodium Carbonate for Activating Slag and Fly Ash Mixtures, Journal of Cleaner Production, vol. 113, pp. 66-75.
43. T. Phoo-ngernkham, A. Maegawa, N. Mishima, S. Hatanaka, and P. Chindaprasirt, (**2015**), Effects of Sodium Hydroxide and Sodium Silicate Solutions on Compressive and Shear Bond Strengths of FA-GBFS Geopolymer, Construction and Building Materials, Vol. 91, pp. 1-8.
44. K. Sankar, P. Stynoski, G.K. Al-Chaar, and W.M. Kriven, (**2018**), Sodium Silicate Activated Slag-Fly Ash Binders: Part I-Processing, Microstructure, and Mechanical Properties, Journal of the American Ceramic Society, Vol. 101, pp. 2228–2244.
45. G. Yu and H. Wang, (**2018**), Study on the Performance of Alkali-Based Geopolymer under Vacuum and Freeze-Thaw Cycles, Chemical Engineering Transactions, Vol. 71, pp. 841-846.
46. J. Qiu, Y. Zhao, J. Xing, and X. Sun, (**2019**), Fly Ash/Blast Furnace Slag-Based Geopolymer as a Potential Binder for Mine Backfilling: Effect of Binder

Type and Activator Concentration, Advances in Materials Science and Engineering, Vol. 2019, pp. 1-13 (https://doi.org/10.1155/2019/2028109).
47. G.S. Mallika, S. Somasundaram, and K.A. Ramaswamy, (**2020**), A Review on Recent Development in Geopolymer Composites, AIP Conference Proceedings, Vol. 2240, pp.090005-xxx (https://doi.org/10.1063/5.0011080).
48. A.E. Kurtoğlu, R. Alzeebaree, O. Aljumaili, A. Ni, M.E. Gülşan, G. Humur, and A. Çevik, (**2018**), Mechanical and Durability Properties of Fly Ash and Slag based Geopolymer Concrete, Advances in Concrete Construction, Vol. 6, pp. 345-362 (https://doi.org/10.12989/acc.2018.6.4.345).
49. Y.-C. Ding, T.-W. Cheng, and Y.-S. Dai, (2017), Application of Geopolymer Paste for Concrete, Repair, Structural Concrete, Vol. 18, pp. xxx-xxx (https://doi.org/10.1002/suco.201600161).
50. Sindhunata, J.S.J. Van Deventer, G.C. Lukey, and H. Xu, (**2006**), Effect of Curing Temperature and Silicate Concentration on Fly-Ash-Based Geopolymerization, Ind. Eng. Chem. Res., Vol. 45, pp. 3559-3568 (https://doi.org/10.1021/ie051251p).
51. J. Xie, J. Zhao, J. Wang, C. Wang, P. Huang and C. Fang, (2019), Sulfate Resistance of Recycled Aggregate Concrete with GGBS and Fly Ash-Based Geopolymer, Materials, Vol. 12, pp. 1247-18 (DOI:10.3390/ma12081247).
52. M.B. Karakoc, I. Turkmen, M.M. Maras, F. Kantarci, R. Demirbog, and M.U. Toprak, (**2014**), Mechanical Properties and Setting Time of Ferrochrome Slag Based Geopolymer Paste and Mortar, Construction and Building Materials, Vol. 72, pp. 283-292.
53. S. Jena and R. Panigrahi, (**2019**), Performance Assessment of Geopolymer Concrete with Partial Replacement of Ferrochrome Slag as Coarse Aggregate, Construction and Building Materials, Vol. 220, pp. 525-537.
54. T. Falayi, (**2019**), Sustainable Solidification of Ferrochrome Slag through Geopolymerization: A Look at The Effect of Curing Time, Type of Activator and Liquid Solid Ratio, Sustainable Environment Research, Vol. 29, pp.1-10.
55. M. Panigrahi, S. Mohanty, R.R. Dash, and R.I. Ganguly, (**2018**), Development of Novel Constructional Material from Industrial Solid Waste as Geopolymer, IOP Conf. Series: Materials Science and Engineering, Vol. 410, pp. 1-12 (DOI:10.1088/1757-899X/410/1/012012).
56. M. Panigrahi, S. Mohanty, R.I. Ganguly, and R.R. Dash, Synthesis and Characterizations Of High Carbon Ferrochrome (HCFC) Slag Based Geopolymer, IntechOpen Publisher, UK, pp. 1-16, SBN: 978-1-83968-989-5 (DOI: http://dx.doi.org/10.5772/intechopen.97140).

6

Pond Ash (PA)–Jute Fiber-Based Geopolymer Cementitious Materials

Muktikanta Panigrahi[1]*, Paresh Biswal[2], Niharika Patel[2], Ratan Indu Ganguly[3] and Radha Raman Dash[4]

[1]PG Department of Materials Science, Maharaja Sriram Chandra Bhanja Deo University, Keonjhar Campus, Odisha, India
[2]Department of Civil Engineering, Gandhi Institute of Engineering and Technology, University, Gunupur, Odisha, India
[3]Department of Metallurgical Engineering, National Institute of Technology, Raurkela, Odisha, India
[4]CSIR-National Metallurgical Laboratory, Jamshedpur, Jharkhand, India

Abstract

Due to its cost-effectiveness and ecological performance, geopolymer (GP) reinforced with natural fibers has the potential to overcome the inherent deficiencies in cementitious materials. Use of natural fibers, especially jute fiber, in a relatively brittle cementitious matrix has achieved considerable strength and toughness in the composite. Durability of such fibers in a highly alkaline matrix must be taken into consideration through effective modifications. Jute fiber (0.5%) has been used as reinforcement in cementitious composite, i.e., GP/jute fiber composite, GP/jute fiber mortar and GP/jute fiber concrete. The analysis presented in this chapter shows the strengthening performance of GP/jute fiber-based materials, with the maximum strength of GP with jute fiber, GP mortar with jute fiber, and GP concrete with jute fiber being 16.45 MPa, 22 MPa, and 28.96 MPa, respectively, at optimum treatment combinations. Prepared products are also characterized by SEM, FTIR, and TGA, with the TGA patterns revealing the striking features observed for three materials (GP with jute fiber, GP mortar with jute fiber, and GP concrete with jute fiber). Finally, the performance of the materials (GP mortar with jute fiber and GP concrete with jute fiber) is evaluated at different test temperatures (70 °C, 80 °C, and 90 °C).

**Corresponding author*: muktikanta2@gmail.com

Muktikanta Panigrahi, Ratan Indu Ganguly and Radha Raman Dash (eds.) Development of Geopolymer from Pond Ash-Thermal Power Plant Waste: Novel Constructional Materials for Civil Engineers, (169–194) © 2023 Scrivener Publishing LLC

***Keywords*:** Pond ash, jute fiber, geopolymer, compressive strength, TGA

6.1 Introduction

Nowadays, fiber-reinforced composite materials (especially treated with alkali-activated materials) play an important role in finding many technical applications in aerospace, automotive, naval architecture and ground transportation [1, 2]. Uses of composite materials have plenty of technical applications over traditional materials. Technological development has enabled efficient arrangements of fibers in the matrix with improved mechanical properties, particularly fracture toughness [3]. The fiber of composite materials overcomes cracking problems and brittle nature, enhancing their ductile behavior [4]. Therefore, damage caused by cracking may be mitigated. Another goal for using fiber as reinforcement is to increase flexural strength of composites [5, 6].

In view of the above, it is thought to be beneficial to develop geopolymer from pond ash. It is a challenge to develop new construction materials, i.e., geopolymer from pond ash, fly ash, and red mud.

Industries like NALCO, BALCO, and NTPC, etc., are major producers of pond ash. Pond ashes are considered to be pollutants. Dumping and landfilling of this waste is a major issue from a land utilization point of view. Additionally, other industrial wastes, such as fly ash, red mud and other industrial slags, are dumped into water-containing ponds to reduce air pollution. Hence, researchers are now showing interest in effective utilization of the above wastes. It is felt that interdisciplinary research will be a meaningful solution to the above problems.

In the present work, geopolymer composite is developed utilizing pond ash and natural fibers. The most common fibers used in geopolymer matrix are fibers (carbon or glass fibers [8, 10, 11], natural fibers obtained from different plants, *viz*., banana, bamboo, flax, and jute). Natural fibers are not only eco-friendly but also easily available and cost-effective. The important properties gained are lower density and enhanced thermomechanical properties in comparison to geopolymer prepared with synthetic fibers. Jute fibers add other properties, such as higher tenacity, lower bulk density, and better heat insulation, thereby making value-added composite for applications [12].

Reinforcing traditional cement and cementitious materials with natural fibers has become popular. However, only a few studies are focused on reinforcement of geopolymer-based composites with natural fibers. A few researchers have studied geopolymer composite using various natural fibers such as sweet sorghum [7], cotton [8, 9], wool [13] and cellulose [14], peat wood [15], wood flour [16] or sawdust [17]. Promising results were observed.

Bera *et al.* [18] studied shear strength of reinforced pond ash. They conducted a series of undrained unconsolidated (UU) triaxial tests on both unreinforced as well as reinforced pond ash. Also, types of reinforcing fibers on shear strength of pond ash-based materials were studied.

Korniejenko *et al.* [20] prepared geopolymer composites using different short natural fibers (i.e., cotton, sisal, raffia and coconut). They evaluated the mechanical properties (compressive strength and flexural strength) of the geopolymer composite. The microstructural behavior provides a better understanding of the strengthening mechanism. Addition of an appropriate amount of natural fiber to geopolymer composites will improve mechanical properties.

Alomayri *et al.* [21] synthesized geopolymer composites with cotton fiber as reinforcement with varying fiber content (from 0–1.0 wt%). After characterizing hardness, impact strength and compressive strength, they concluded that an optimum weight fraction of fiber content (0.5 wt%) of the composites gave the best properties.

Ribeiro *et al.* [22] prepared geopolymer composite reinforced with bamboo fibers and processed bamboo fibers (mixed potassium-sodium polysialate siloxo-type geopolymer reinforced with bamboo fibers and strips). They found higher strength in processed bamboo fibers. Also, they synthesized processed Amazonian kaolin (i.e., metakaolin-based geopolymer). Different techniques (i.e., X-ray diffraction, scanning electron microscopy (SEM), four-point flexural and compressive strength testing) were employed to determine the structure, morphology, and strength properties of as-prepared composite.

Agopyan *et al.* [23] prepared asbestos in roofing components. In the preparation of roofing components, they used brittle materials like cement as matrix and different fibers, such as vegetable fibers, freshly prepared pulp from eucalyptus waste, residual sisal and coir fibers, as reinforcing components. They evaluated mechanical and durability properties, which were improved. They also used an alternative binder (*viz.*, free lime using ground granulated blast furnace slag). This composite showed better performance.

Malenab *et al.* [24] prepared geopolymer composite using natural fibers as eco-friendly reinforcement components. Also, they used treated waste fibers in the preparation of eco-friendly geopolymer composite. Treated fiber-based eco-friendly geopolymer composite showed improved adhesion property. They optimized NaOH pretreatment parameters such as soaking time in aluminium salt solution and pH of slurry. It was observed that the properties and performance of pretreated geopolymer composite was proven to be better in comparison to untreated geopolymer composite.

He *et al.* [25] prepared geopolymer composite using red mud (RM) and rice husk ash (RHA) in different proportions. Prepared composite was characterized with compression test, X-ray diffraction, and scanning electron microscopy analyses; and its physico-mechanical properties were evaluated. They observed that there was a reduction of ductility in materials when compared with red mud-based geopolymer. With higher RHA/RM ratios, strength (stiffness and ductility) of the composite improved. Microstructural and compositional analyses were done on final products.

Detphan and Chindaprasirt *et al.* [26] prepared different batches of geopolymer using different ratios of rice hush ash and fly ash in different proportions (0/100, 20/80, 40/60, and 60/40). Geopolymer mortar was prepared mixing geopolymer with sand. Upon investigating the effect of RHA fineness on density and strength properties of geopolymer mortars, they observed compressive strength ranging between 12.5 to 56.0 MPa for different mass ratios of RHA/FA. They concluded that the strength property mainly depends on FA/RHA, RHA fineness, and ratio of sodium silicate to NaOH. For the highest strength, optimum parameters of FA-RHA geopolymer mortars were found to be 4.0 sodium silicate/NaOH mass ratios, delay time before subjecting the samples to heat for 1 h, curing temperature (60 °C), and curing time (48 h).

Younes *et al.* [27] blended recycled or reused waste materials for production of ternary cement mortars (TBCMs) by partial substitution of ordinary Portland cement (OPC) with 20% waste glass powder (WG) and cement (80% OPC). They also used ternary blending with 2.5–10% rice husk ash replacing proportionate amounts of WG. The mortars were cured under tap water for different lengths of time, namely 3/7/28/60/90 days. The results indicated that glass and rice husk ash helped to improve compressive strength with increasing hydration time. Physical parameters, such as overall porosity and percentage water absorption, were simultaneously decreased. The compressive strength results of TBCMs indicated higher value than CM. The impact of γ-radiation on physico-mechanical properties of unsaturated polyester (UP)/impregnated blended cement mortar composite (TBCMs) was studied. Data showed enhancement of mechanical properties for irradiated specimens when compared with unirradiated ones. Furthermore, thermogravimetric analyses (TGA) and XRD analyses were explored to correlate the structure and properties of new materials.

Rill *et al.* [28] used basalt strands (hacked and fiber waves) for making geopolymer composite. They studied the correlation of mechanical behavior and microstructure of prepared composites.

Camargo *et al.* [29] prepared cement and geopolymer-based composite using natural fibers as reinforcing components. Reinforced fibers are renewable, biodegradable, and nontoxic in nature, and exhibit attractive mechanical properties (compared with their synthetic fiber counterparts). Their study on wet behavior and the impact of modified and functionalized fiber on durability showed substantial improvement in mechanical properties.

Zulfiati *et al.* [30] prepared geopolymer mortar and concrete using fly ash as matrix and different percentages (0%, 0.25%, 0.50%) of pineapple fiber as reinforcement. By experimenting with different combinations of reinforcement, they observed the attainment of compressive strength of 41.468 MPa and flexural strength of 9.209 MPa using maximum fiber length of 30 mm, 0.50% volume fraction and 16 M NaOH, respectively.

Wattanasiriwech *et al.* [31] prepared new generation of geopolymer composite using bagasse cellulose fiber as reinforcing component. The prepared materials promise good thermal stability and chemical resistance with adequate mechanical properties. Other important improved properties of the composites were impact strength, wettability and density. It was reported that additions of bagasse cellulose fibers up to 1–2% improved impact strength of geopolymers. Any excess addition of fibers, however, resulted in the decrease of impact strength. This is due to agglomeration of fibers with lesser degree of intimation with matrix. Hydrophilicity of these fibers leads to an increase of water absorption with decrease in density of composites.

Sarmin [32] used fly ash (Class F) and metakaolin (30%) as base materials for preparing lightweight geopolymer composites. During the preparation of composite, they used wood particles (WP), wood flour (C100), and wood fiber (WF) as reinforcement. Geopolymer reaction was augmented by treating with alkaline activator solution. Mechanical and other properties of geopolymer composites were evaluated with different treatment combinations, i.e., curing times (20 °C for 7, 14, and 28 days and 80 °C for 6 and 24 h). The highest compressive strength was found to be 38.4 MPa with wood flour cured at 20 °C for 28 days.

Elavarasan and Dhanadeepa [33] prepared geopolymer concrete using 90% rice husk ash (RHA) and 10% cement with different fibers by using alkaline liquids. During the composite preparation, glass fibers (volume fractions of 0.03%) and different volume fractions of polypropylene fibers (0.25%, 0.5%, 0.75%) were used. Strength of geopolymer composite is enhanced due to the presence of amorphous silica (i.e., cellular form). Another advantage of RHA is that it created impermeability, and thus improved durability. RHA minimized alkali-aggregate reaction and reduced the expansion of concrete. For geopolymer reaction, alkaline liquid to rice husk ash ratio was maintained at 0.4. Quantitative relationship was developed between strength properties (tensile and compressive) and volume fraction of glass fiber/polypropylene fiber by regression analysis.

Ohno and Li [34] developed strain-hardening ductile fiber-reinforced fly ash-based geopolymer composites by using randomly oriented short poly-vinyl alcohol (PVA) fibers. Their mechanical properties were investigated by cube compressive and dog bone tensile testing. Tensile strain hardening behavior with high ductility (over 4%) was experimentally demonstrated for the composites. The performances of cured samples were found to be better than uncured samples. Crack width distributions were also investigated by using the digital image correlation technique. The maximum and average crack widths were found to be 117 μm and 45 μm, respectively, even at a high imposed strain level (4.5%), thereby establishing the feasibility of strain-hardening ductile geopolymer composites.

Rangan *et al.* [35] prepared geopolymer foams using aluminum powder and several combinations of silica fume and fine sand reinforced with short basalt fibers. Curing was done at room temperature and then the samples were put in a furnace at 70 °C for curing time of 7 and 28 days. They evaluated thermal conductivity, porosity, compressive strength, flexural strength and density. All the test results were found to be 0.13–0.359 $W{\cdot}m^{-1}{\cdot}K^{-1}$, 41.8–62.5%, 1.94–9 MPa, 0.96–2.93 MPa, 546–1028 $kg{\cdot}m^{-3}$. Results indicated that filler had a significant effect on the mechanical and physical properties of the tested samples.

Subramani and Sakthivel [36] studied as-prepared eco-friendly fly ash-based geopolymer bricks (size = 190 mm × 90 mm × 90 mm). They used different ratios of river sand and silica sand (1:1.6, 1:1.8, 1:2) by weight. Cast bricks were cured at ambient temperature and then exposed for 18 weeks in H_2SO_4 media. It was observed that the bricks did not loose their alkalinity. The weight loss of bricks was observed to range from 0.54% to 0.28% of their initial weight value. It was also observed that weight loss was higher when bricks were made with geopolymer prepared with a higher percent of Na_2SiO_3. Results indicated that geopolymer bricks are highly resistant to sulfuric acid.

Shaikh and Haque [37] synthesized geopolymer using fly ash and potassium activators. They used carbon and basalt fiber as reinforcing component in a geopolymer mold in order to make a composite. Curing is done at both ambient and elevated temperatures. For comparison purpose, they prepared geopolymer (blank) and geopolymer composite with fiber. Six series of fly ash-based geopolymer were made using different weight percent (wt%) of carbon and basalt fiber (0.5%, 1%, 1.5%). Properties of geopolymer were found at ambient temperature and higher temperature (200, 400, 600 and 800 °C). Better properties, i.e., compressive strength, were found for the composite prepared with 1 wt% basalt and 1 wt% carbon fiber. Less volumetric shrinkage and mass loss were also observed. Microstructure of carbon and basalt fiber-reinforced geopolymer composite was correlated with strength properties.

The previous reviews have helped the present authors define the present work, i.e., preparation of geopolymer-based composite from power plant waste, i.e., pond ash. In the present work, instead of fly ash, pond ash was chosen for preparation of geopolymer. Jute fiber was used as reinforcing agent in the preparation of geopolymer composite; and activated alkaline solution was used to prepare geopolymer. The strength property of geopolymer composite was established by measuring compressive strength. Prepared materials were characterized by X-ray diffraction to identify different phases in the materials. FTIR characterization was done to find various absorption bands present in the prepared materials. Thermal stability was determined using TG-DTG analysis; and DSC analysis was carried out to study the geopolymerization process. Surface topological analyses were done using SEM.

6.2 Experimental Details

6.2.1 Chemicals and Materials

Pond Ash, the main ingredient for geopolymer, was collected from NALCO Damanjodi, Odisha. Jute fiber, sand and aggregates were obtained locally. Sodium hydroxide (activator) was obtained from M/S Loba Chemicals, India. Sodium silicate (Na_2SiO_3) was procured from Merck, India. Water-soluble plasticizer (Sika) was purchased from a reputed firm located in the southern part of India (Visakhapatnam market). Plasticizer usually enhances plasticity of the products, i.e., geopolymer/geopolymer mortar/concrete.

Natural fibers are considered superior alternatives to synthetic fibers due to their high flexural modulus and impact strength. Additionally, they are environmentally friendly, abundantly available, renewable, and are low density and cost-effective. The natural fibers can contribute to a healthy ecosystem, while their low cost and high performance are economically beneficial to industries [37].

Jute is a natural fiber which can be used for sacking, burlap, and twine as a backing material for tufted carpets. It's a long, soft, shiny bast fiber that

Figure 6.1 Photograph of jute fibers.

can be spun into coarse, strong threads. It is one of the most affordable natural fibers and second only to cotton in the amount produced and variety of uses.

Jute fibers are primarily composed of plant materials (i.e., polyose, lignin, and pectin). A photograph of jute is shown in Figure 6.1 and the physical properties of jute fiber are given in Table 6.1.

6.2.1.1 Physical Properties of Jute Fiber

The physical properties of jute fiber(s) are presented in Table 6.1 [26, 38] and discussed below.

A brief overview of the properties and uses of jute fiber follows:

- Jute fiber is 100% biodegradable and recyclable, and consequently earth benevolent.
- Jute fiber is characteristically golden and has a silky shine, hence it is called The Golden Fiber.
- Jute is the least expensive vegetable fiber secured from the bast or skin of the plant's stem.
- Jute is the second most important vegetable fiber after cotton as far its use, global consumption, production, and availability.
- Due to its high tensile strength and low extensibility, better breathability of fabrics is ensured. Subsequently, jute is very suitable for agricultural commodity bulk packaging.

Table 6.1 Physical properties of jute fiber.

Sl. no.	Parameters	Values
1	Tenacity (g/den)	3.5–4.5 (40–70g/tex)
2	Length	0.2–30 inch
3	Stretch and Elasticity	Not good and 2% elongation at break
4	Resiliency	Not very good
5	Abrasion Resistance	Relatively good
6	Dimensional Stability	Good
7	Moisture Regain	13.75%
8	Specific Gravity	1.48–1.50
9	Color	Yellowish, yellow, brown, golden
10	Heat Resistance	Good
11	Specific Heat	0.324
12	Diameter	18 microns
13	Cross Section	Uneven, thick cell wall with lumen

- Jute makes the best quality industrial yarn, fabric, net, and sacks. It is a standout amongst the most adaptable natural fibers that have been utilized in raw materials for packaging, textiles, non-textile, construction, and agricultural sectors. Yarn bulking results in less breaking tenacity and increased breaking extensibility when blended as a ternary blend.
- Unlike hemp, jute is not a form of cannabis, and therefore is more easily distinguished from forms of cannabis that produce a narcotic.
- Jute is a standout amongst the most versatile natural fibers utilized in raw materials for packaging, textiles, non-textile, and agricultural sectors.
- Jute stem has a high volume of cellulose that can be procured within 4–6 months, and therefore it can save the forest and meet the cellulose and wood requirement of the world.

Table 6.2 Chemical composition of jute fiber(s).

Components	Percentage
Cellulose	65.2
Hemicellulose	22.2
Lignin	10.8
Water	1.5
Fat and Wax	0.3

- The best varieties of jute are Bangla Tosha – *Corchorus olitorius* (Golden shine); and Bangla White – *Corchorus capsularis* (Whitish Shine). And Mesta or Kenaf (*Hibiscus cannabinus*) is another species with fiber like jute with mid-range quality.
- Raw jute and jute merchandise are translated as burlap, industrial hemp, and kenaf in a few sections of the world.
- The best wellspring of jute on the planet is the Bengal Delta Plain, which is occupied by Bangladesh and India.

Thus, jute is the most environmentally friendly fiber, from seed to expired fiber, as the expired fibers can be recycled more than once. The percentage of chemical composition of jute fiber is presented in Table 6.2.

6.2.2 PA/Jute Fiber-Based Geopolymer, Mortar and Concrete

Synthesis of pond ash-based geopolymer is achievable through the solid-state route. During the preparation of PA/Jute fiber-based geopolymer, pond ash (as matrix components) and jute fiber (as reinforcement components) are used and activated through an alkali activator. The following steps are adopted to obtain PA/Jute fiber-based geopolymer [39, 40].

Step 1: Raw Material Preparation

The as-received pond ash is dried at 120 °C for 2 h in a heating oven. Dried pond ash is crushed or ground followed by sieving through 240 meshes. The pond is called processed pond ash. Pond ash-based geopolymer is prepared by taking an appropriate proportion of pond ash, chopped fiber

Table 6.3 PA/Jute fiber-based GP with same compositions.

Sample code	PA (84.15 g) + jute fiber (0.85 g)	Sod. silicate (SS)	Alkali (8 M, NaOH) (SH)	Water-soluble plasticizer (Sika)
S1	85%	12%	3%	1–2 ml
S2	85%	12%	3%	1–2 ml
S3	85%	12%	3%	1–2 ml

(small length fibers obtained by cutting long fibers aided by a cutter), sodium silicate, sodium hydroxide, and water-soluble plasticizers.

Step 2: Pond Ash/Jute Fiber Geopolymer Preparation

Pond ash is mixed with chopped jute fiber, NaOH, Na_2SiO_3, a small quantity of water (6%) and water-soluble plasticizers (Sika); and a gel-like substance is fashioned. PA/Jute fiber-based GP with same compositions is shown in Table 6.3. Three samples are designated as S1, S2, and S3, respectively.

Step 3: Molding, Casting and Compaction

As-prepared gel-like materials are cast in a cylindrical REMI iron mold having dimensions of 50 mm diameter and 70 mm height. Samples are then rammed twenty times to enable better compaction. Then they are left in the mold for 3 min. Samples are finally demolded to obtain cylindrical green samples.

Step 4: Curing Process

Demolded samples are cured at 70 °C for 24 h in an oven. These samples are now ready for different characterizations.

PA/Jute Fiber Mortar Preparation:

Pond ash/jute fibers-based geopolymer mortar is prepared by chemical approach through the solid-state route. It is prepared by mixing a proportionate amount of PA, jute fiber and sieved sand (see Table 6.2). The preparation of PA/jute fiber-based geopolymer mortar is described below [39, 40]:

Step 1: Raw Material Preparation

The collected pond ash is dried at 120 °C in a heating oven. The dried ash is ground and sieved through 240 meshes, called process pond ash. Collected sand is sieved and dried at 120 °C for 2 h in a heating oven, called process sand. Jute fiber is cut into small pieces of fibers and is known as chopped fibers. Chopped fibers are used to prepare PA/Jute fiber-based GP mortar.

Sodium silicate (Na_2SiO_3) and sodium hydroxide (NaOH) are combined to prepare an alkaline solution used as geopolymer activator. Water-soluble plasticizer is used to increase plasticity of geopolymer mortar.

Step 2: Geopolymer Mortar Preparation

The processed pond ash is mixed with chopped fibers, NaOH, Na_2SiO_3, a small quantity of water (6%) and water-soluble plasticizers (Sika). This will result in the formation of a gel-like substance. The gel-like substance, i.e., GP is mixed with a requisite amount of sand to form GP mortar (for compositions see Table 6.2).

Step 3: Molding, Casting and Compaction

The gel-like materials are placed in a REMI mold made up of iron having a cylindrical shape (50 mm diameter size and 70 mm height). Materials are rammed 20 times and then they are left for 3 min before demolding. Demolded samples are known as "green" mortar samples.

Step 4: Curing Process

Green mortar samples are cured at 70 °C for 24 h in a heating oven. The cured samples are then sealed in plastic zipper bags to prevent moisture absorption till they are tested. Compositions of PA/Jute fiber-based mortar are shown in Table 6.4.

The PA/Jute fiber-based concrete preparation process is accomplished by a chemical approach through solid-state route. The successive steps of PA/Jute fiber-based concrete preparation are grinding, mixing, ramming and curing (using a heating oven) [39, 40].

Table 6.4 Compositions of PA/Jute fiber-based mortar mixtures (Mortar = 1.0:3.0).

Sample code (Three sets)	PA/Jute fiber (21.125%) + sand (63.375%)	Sod. silicate (SS)	Alkali (8 M, NaOH) (SH)	Water-soluble plasticizer (Sika)
S1	85%	12%	3%	1–2 ml
S2	85%	12%	3%	1–2 ml
S3	85%	12%	3%	1–2 ml

Step 1: Raw Material Preparation (PA/Jute Fiber-Based Concrete)
As-received pond ash is dried at 120 °C for 2 h in a heating oven. Once dried, it is ground and sieved through 240 meshes. Such pond ash is known as processed pond ash. Procured jute fibers are cut into small pieces. Small-sized fibers are called chopped fibers. Collected sand is sieved and dried at 120 °C for 2 h in a heating oven. This is known as processed sand. Stone chips for making GP concrete are dried at 120 °C for 1 h. All the components are kept separately. Geopolymer activator is prepared by mixing Sodium silicate and sodium hydroxide in appropriate proportions. Water-soluble plasticizers are used to enhance the plasticity of PA/Jute fiber-based concrete.
Step 2: PA/Jute Fiber-Based Geopolymer Concrete Preparation
Processed Pond Ash is mixed with chopped jute fibers, alkaline activator (NaOH and Na_2SiO_3), a small quantity of water (6%) and water-soluble plasticizers (Sika). A gel-like substance is formed from this mixture. The gel-like substance is further mixed with a requisite amount of sand and stone chips (aggregates) to form GP concrete (for composition see Table 6.5). Three replicates are made for checking reproducibility of the findings and are designated as S1, S2, and S3, respectively.
Step 3: Molding, Casting and Compaction
Jute fiber GP concrete samples are poured into a cylindrical shape REMI mold made of iron with dimensions of 50 mm diameter and 70 mm height. The material inside the mold is rammed 20 times to increase the compactness of the concrete. It is left for 3 min and then is demolded. This sample is known as "green" sample.
Step 4: Curing Process
Green samples are cured at 70 °C for 24 h in a heating oven. Samples are preserved in plastic zipper bags to prevent moisture absorption.

Table 6.5 Compositions of PA/Jute fiber-based concrete mixtures (Concrete = 1.0:1.5:3.0).

Sample code (Three sets)	PA/Jute fiber (15.36364%) + sand (23.04546%) + aggregates (46.09092%)	Sod. silicate (SS)	Alkali (8 M, NaOH) (SH)	Water-soluble plasticizer (Sika)
S1	85%	12%	3%	1–2 ml
S2	85%	12%	3%	1–2 ml
S3	85%	12%	3%	1–2 ml

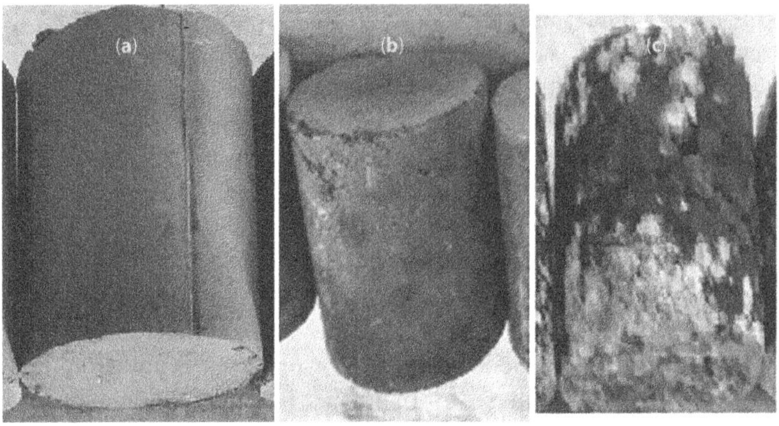

Figure 6.2 Photographs of (a) PA/jute fiber-based GP, (b) GP mortar, and (c) GP concrete.

Photographs of PA/jute fiber-based GP, PA/jute fiber-based GP mortar, and concrete are shown in Figure 6.2.

Test Methods for Pa/Jute Fiber-Based Geopolymer:
Compressive strength tests were performed on cylindrical specimens (50 mm × 70 mm) using a 20 KN digital compressive testing machine (Mechatronic Control System; Model: MECH-CS/UTE 20T). Experiments were replicated in batches to ascertain validity of test results. Compression tests were performed at three temperatures (70 °C, 80 °C, 90 °C). Test results ensured the integrity of GP material at higher temperatures (70 °C, 80 °C, 90 °C).

Studies with a scanning electron microscope (SEM) attached with an energy dispersive X-ray (EDX) system helped in understanding the morphology and chemical compositions of prepared materials.

Samples were further characterized with FTIR (Thermo Nicolet Nexus 870 spectrophotometer). The settings for the instruments are kept constant (50 scan at 4 cm^{-1} resolution, absorbance mode) during the entire examination.

Thermal stability of the PA/Jute fiber-based GP and GP mortar/concrete was determined by thermogravimetric analysis.

Enthalpy of fusion for as-received PA/Jute fiber-based GP and GP mortar/concrete was estimated using differential scanning calorimetry (DSC). The parameters, such as heating rate, temperature range and nitrogen environment (N2), were set as 10 °C/min and 50–300 °C, respectively.

Table 6.6 Compressive strength of Pond Ash–Jute Fiber-based geopolymer materials.

S. no.	PA	JF	Sand	Aggregate	SS (%)	SH (%)	H_20 (ml)	SIKA (ml)	TEMP. (°C)	CS (MPa)	Av. CS (MPa)
(A) Compressive strength of pond ash–jute fiber-based geopolymer											
S1										17	
S2	84.5	0.5	---	---	12	3	6	1-2	70 16.3 17.1		16.45
S3											
(B) Compressive strength of pond ash–jute fiber-based geopolymer mortar											
S1					12	3	6	1-2	70	22.3	22.0
S2	21.125	0.5	63.375	---						22.1	
S3										21.8	
(C) Compressive strength of pond ash–jute fiber-based geopolymer concrete											
S1										28.8	
S2	15.36364	0.5	23.04546	46.09092	12	3	6	1-2	70	29.2	28.96
S3										28.9	

Note: PA = Pond Ash, JF = Jute Fiber, SS = Sodium Silicate, SH = Sodium hydroxide, H_20 = Water, TEMP = Temperature, CS = Compressive Strength, Av. = Average, MPa = Mega Pascal.

Table 6.7 Compressive strength of PA-based GP materials (GP mortar and GP concrete).

(A) Compressive strength of PA-based GP mortar at temp. 70 °C, 80 °C, 90 °C [SS (12%), SH (3 %, 8 M), H_20 (6 ml), Sika (1-2 ml)]									
			Components (85%)						
S. no.	Sample name	Replication	PA	JF	FA	GA	TM (°C)	CS (MPa)	Av. CS (MPa)
1	PA/Jute fiber GP Mortar	S1	21.125	0.5	63.375	---	70	22.3	22.0
		S2						22.1	
		S3						21.8	
2	PA/Jute fiber GP Mortar	S1	21.125	0.5	63.375	---	80	21.8	22.3
		S2						22.5	
		S3						22.4	
3	PA/Jute fiber GP Mortar	S1	21.125	0.5	63.375	---	90	22.2	20.9
		S2						20.8	
		S3						20.9	

(Continued)

Table 6.7 Compressive strength of PA-based GP materials (GP mortar and GP concrete). *(Continued)*

(B) Compressive strength of PA/Jute fiber-based GP concrete at temp. 70 °C, 80 °C, 90 °C [SS (12%), SH (3 %, 8 M), $H_2$0 (6 ml), Sika (1-2 wml)]									
1	PA/Jute fiber GP Concrete	S1	15.36364	0.5	23.04546	46.09092	70	28.8	28.96
		S2						29.2	
		S3						28.9	
2	PA/Jute fiber GP Concrete	S1	15.36364	0.5	23.04546	46.09092	80	29.5	29.4
		S2						29.5	
		S3						29.3	
3	PA/Jute fiber GP Concrete	S1	15.36364	0.5	23.04546	46.09092	90	28.7	28.7
		S2						28.8	
		S3						28.5	

*Note: PA (Pond Ash), GP (Geopolymer), JF (Jute Fiber), Fine aggregate (FA), Granule aggregates (GA), TM (Testing temperature), CS (Compressive Strength), Av. (Average), MPa (Mega Pascal).

6.2.3 Results and Discussion

Table 6.6A, 6.6B, and 6.6C describe the compressive strength of PA/Jute fiber geopolymer (GP), PA/Jute fiber GP mortar, and PA/Jute fiber GP concrete, respectively. For each category of materials, replication is made in order to obtain average (av.) strength of materials. The highest strength is observed for PA/Jute fiber GP concrete. The enhancement of strength of PA/Jute fiber GP mortar and PA/Jute fiber GP concrete is thought to be due to formation of higher degree of chain [41–43].

Tables 6.7A and 6.7B describe the compressive strength values of GP mortar and GP concrete at three temperatures, i.e., 70 °C, 80 °C, 90 °C, respectively. For accuracy, replicates are made for each batch (S_1, S_2, S_3), where subscripts 1, 2, 3 represent samples from each batch. The purpose of conducting the test at higher temperatures of the two materials is to discover the relative stability of materials at higher temperatures. Results suggest materials retained their stability at higher temperature.

Microstructures of PA/Jute fiber GP, PA/Jute fiber GP mortar, and PA/Jute fiber GP concrete were analyzed with SEM (Figure 6.3). General features of the micrographs indicate the presence of compact mass, presumably of glassy phases formed due to polymerization of the materials occurring during treatment with alkali solutions. Chopped fibers of different lengths and diameters (i.e., different aspect ratio) are dispersed in the matrix. The presence of jute fiber enhances the mechanical property of the materials.

Superimposed FTIR spectra of four materials, i.e., Jute fibers, PA/Jute fiber-based GP, PA/Jute fiber-based GP-Mortar, and PA/Jute fiber-based GP-Concrete are shown in Figure 6.4. The FTIR spectrum of Jute fiber shows different peaks at 2800–3000, 2952 and 2872 cm^{-1}, respectively. CH_3

Figure 6.3 SEM images of (a) PA/Jute fiber-based GP, (b) PA/Jute fiber-based GP mortar, and (c) PA/Jute fiber-based GP concrete.

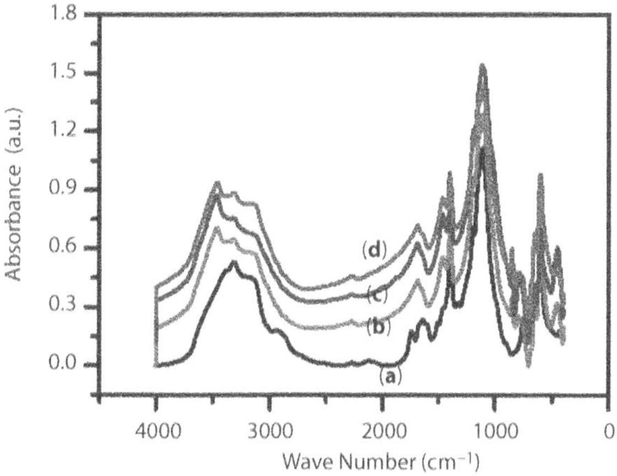

Figure 6.4 FTIR spectra of (a) Jute fiber, (b) PA/Jute fiber-based GP, (c) PA/Jute fiber-based GP-Mortar, and (d) PA/Jute fiber-based GP-Concrete.

asymmetric and symmetric stretching vibration is allocated at 2800–3000 cm^{-1}. Peaks at 2920 and 2839 cm^{-1} are ascribed to CH_2 asymmetric and symmetric stretching vibrations. Spectrum is also found at 1456 and 1375 cm^{-1}. Peak at 1456 cm^{-1} is attributed to CH_3 asymmetric deformation vibrations (or CH_2 scissor vibrations). Peak at 1375 cm^{-1} is attributed to symmetric deformation vibrations of CH_3 group. Peaks at wavenumber range 800–1200 cm^{-1} are attributed to C-C asymmetric stretching, whereas peak at 1160 cm^{-1} is attributed to asymmetric rocking of CH_3. A C-H wagging vibration is found at 998 cm^{-1}. Peak at 972 cm^{-1} is assigned to CH_3 asymmetric rocking and C-C asymmetric stretching vibrations. Peak at 898 cm^{-1} is attributed to CH_3 asymmetric rocking and C-C asymmetric and symmetric stretching vibrations. Peaks at 840 and 808 cm^{-1} are caused by CH_2 rocking vibrations [44, 45].

The FTIR spectrum of PA/Jute fiber GP is shown in Figure 6.4b, in which a broad stretching band can be found around 2919.50 cm^{-1}. The band arises due to the stretching vibrations of O-H bonds and H bending vibrations of H-O-H of interlayer adsorbed H_2O molecule [44–47]. The C=O stretching band is indicated at 1449.40 cm^{-1} and confirms the presence of carbonate groups [44–47]. The main reason for this being the presence of chemisorbed CO_2 in PA/Jute fiber GP. The Si-O and O-Si-O bands are also found at 989.30 cm^{-1} and 534.50 cm^{-1}, which indicates the presence of silicate groups. The presence of Al^{3+}-O^{2-} bonds are also observed near 805.5 cm^{-1}; and an Fe-O stretching band is also observed at 440 cm^{-1}.

The FTIR spectrum of PA/Jute fiber GP mortar is shown in Figure 6.4c. The peak at 1113 cm^{-1} is attributed to an Si–O band in SiO_4 molecules. A weak Si–O symmetrically stretching band is observed at 640 cm^{-1}. Peaks at 995 cm^{-1} and 790 cm^{-1} are assigned to alternating Si–O and Al–O bonds. Al–O stretching vibration is present in AlO_4 tetrahedra. There is shifting of stretching frequencies. It is observed that the band showing asymmetric bending of the O–Si–O and O–Al–O bands shifts to a lower value, confirming the previous finding [48, 49].

The FTIR spectrum of PA/Jute fiber GP concrete is shown in Figure 6.4d, which indicates the presence of different absorption bands occurring at different wavenumbers, i.e., 3441, 2918, 1823, 1638, 1117 and 907 cm^{-1}. The bands are in agreement with stretching vibrations of O-H bonds (3441 cm^{-1} wavenumber) and H-O-H bending vibrations (1638 cm^{-1} wavenumber) of interlayer adsorbed H_2O molecule [48]. The hydroxyl-stretching band of water plays an important role and peak shift of the FTIR spectra is significant. The ensuing absorption band at 1117 cm^{-1} wavenumber is attributed to Si-O band and signifies the occurrence of silicate groups. The presence of $Al^{3+}O^{2-}$ absorption bands are also indicated near 907 cm^{-1} wavenumber [50, 51].

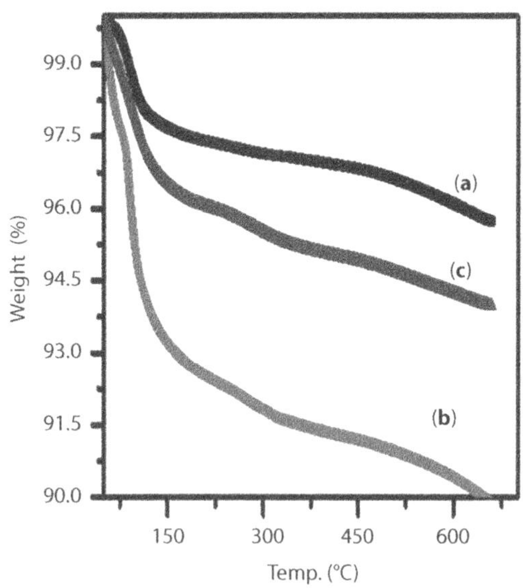

Figure 6.5 TGA plot of (a) PA/Jute fiber-based GP, (b) PA/Jute fiber-based GP-M, and (c) PA/Jute fiber-based GP-C.

Cured samples of as-prepared PA/Jute fiber-based GP materials were charcterized using TG-DTG curves obtained in an isotermal experiment (Figure 6.5). The significant observation made is the loss of water beginning around 50 °C till 400 °C temperature (humps in the thermogram), indicating loss of water since polymerization is completed during curing [52–54].

6.3 Conclusions

In this chapter, extensive studies carried out on pond ash/jute fibers-based geopolymeric materials were presented. Low-calcium content pond ash was used as the source material instead of Portland cement to prepare geopolymeric materials. Pond Ash/Jute Fibers-based geopolymeric materials showed good compressive strength at room temperature as well as at higher temperatures, i.e., 70 °C, 80 °C, 90 °C. Microstructures of jute fiber-based GP composites indicated the occurrence of polymerization in the materials, showing glassy phases with some unreacted materials. Jute fibers with different aspect ratios were observed to be widespread in the matrix.

Acknowledgments

The authors convey their sincere thanks to the Ministry of Mines, Government of India, for providing the grant to carry out this work [Grant number= F.No.:14/54/20214-Met.-IV dated: 29.12.2014]. The authors also convey their sincere thanks to GIET, Gunupur, Rayagada, Odisha, India, for providing lab facilities to do the research work. The authors would also like to thank CRF, IIT Kharagpur, for providing testing facilities.

References

1. F.U.A. Shaikh, (2013) Deflection Hardening Behaviour of Short Fibre Reinforced Fly Ash based Geopolymer Composites, Mater. Design, Vol. 50, pp. 674-682.
2. A. Natali, S. Manzi, M.C. Bignozzi, (2011), Novel Fiber-Reinforced Composite Materials based on Sustainable Geopolymer Matrix, Procedia Engineering, Vol. 21, pp. 1124-1131.
3. S.A. Bernal, J. Bejarano, C. Garzon, R.M. De Gutierrez, S. Delvasto, E.D. Rodriguez, (2012) Performance of Refractory Aluminosilicate Particle/Fiber

Reinforced Geopolymer Composites, Compos. Part B-Eng., Vol. 43, pp. 1919–1928.
4. R. Sakulich, (2011), Reinforced Geopolymer Composites for Enhanced Material Greenness and Durability, Sustainable Cities and Society, Vol. 1, 195-210.
5. K. Korniejenko, J. Mikuła, M. Łach, (2015), Fly ash based Fiber-Reinforced Geopolymer Composites as the Environmental Friendly Alternative To Cementitious Materials, International Conference on Bio-Medical Engineering and Environmental Technology (BMEET-15) March 21–22, London (UK), pp. 208–214.
6. K. Korniejenko, J. Mikuła, and M. Łach, (2015), Characterization of mechanical properties of short glass fiber-reinforced geopolymer composites, ICCST/10: 10th International Conference on Composite Science and Technology, 2-4 September 2015, Lisboan, Portugal.
7. R. Chen, S. Ahmari, L. Zhang, (2014), Utilization of Sweet Sorghum Fiber to Reinforce Fly Ash-based Geopolymer, J. Mater. Sci., Vol. 49, pp. 2548-2558.
8. T. Alomayri, F.U.A. Shaikh, I.M. Low, (2014), Synthesis and Mechanical Properties of Cotton Fabric Reinforced Geopolymer Composites, Compos. Part B-Eng., Vol. 60, pp. 36-42.
9. T. Alomayri, F.U.A. Shaikh, I.M.Low, (2013), Thermal and Mechanical Properties of Cotton Fabric-Reinforced Geopolymer Composites, J. Mater. Sci., Vol. 48, pp. 6746-6752.
10. T. Lin, D. Jia, M. Wang, P. He, D. Liang, (2009), Effects of Fibre Content on Mechanical Properties and Fracture Behaviour of Short Carbon Fibre Reinforced Geopolymer Matrix Composites, B. Mater. Sci., Vol. 32, pp. 77-81.
11. T. Lin, D. Jia, P. He, and M. Wang, (2010), In situ Crack Growth Observation and Fracture Behaviour of Short Carbon Fiber Reinforced Geopolymer Matrix Composites, Mat. Sci. Eng. A-Struct., Vol. 527, pp. 2404-2407.
12. F. Pacheco-Torgal, S. Jalali, (2011), Cementitious Building Materials Reinforced with Vegetable Fibers: A Review, Constr. Build. Mater., Vol. 25, pp. 575-581.
13. M. Alzeer and K.J.D. MacKenzie, (2012), Synthesis and Mechanical Properties of New Fibre-Reinforced Composites of Inorganic Polymers with Natural Wool Fibres, J. Mater. Sci. Vol. 47, pp. 6958-6965.
14. L. Yan, B. Kasal, and L. Huang, (2016), A Review of Recent Research on the Use of Cellulosic Fibres, their Fibre Fabric Reinforced Cementitious, Geo-Polymer and Polymer Composites in Civil Engineering, Compos. Part B-Eng., Vol. 92, pp. 94-132.
15. J. Yliniemi, H. Nugteren, M. Illikainen, M. Tiainen, R. Weststrate, and J. Niinimäki, (2016), Lightweight aggregates produced by granulation of peat wood fly ash with alkali activator, Int. J. Miner. Process., Vol. 149, pp. 42-49.

16. M. Łach, J. Wajs, C. Tempka., K. Korniejenko, and J. Mikuła, Fibre-reinforced Composites with Geopolymer Matrix: Fiber-Geopolymers, in: J. Mikuła, K. Korniejenko (eds.), Innovative, Cost-Effective and Eco-friendly Fibre-based Materials for the Construction Industry, Wydawnictwo Politechniki Krakowskiej, Cracow, 2015, pp. 41-50.
17. P. Duan, C. Yan, W. Zhou, W. Luo, (2016), Fresh Properties, Mechanical Strength and Microstructure of Fly Ash Geopolymer Paste Reinforced with Saw Dust, Constr. Build. Mater., Vol. 111, pp. 600-610.
18. A.K. Bera, A. Ghosh, and A. Ghosh, (2009), Shear Strength Response of Reinforced Pond Ash, Construction and Building Materials, Vol. 23, pp. 2386-2393.
19. A. Sharan, (2011), Strength Characteristics of Fiber Reinforced Pond Ash, pp. 21-27.
20. K. Korniejenko, E. Frączek, E. Pytlak, and, M. Adamski, (2016), Mechanical Properties of Geopolymer Composites Reinforced with Natural Fibers, International Conference on Ecology and new Building materials and products, ICEBMP 2016. Procedia Engineering, Vol. 151, 388-393.
21 T. Alomayri, H. Assaedi, F.U.A. Shaikh, and I.M. Low, (2013), Synthesis and Characterization of Mechanical Properties in Cotton Fiber-Reinforced Geopolymer Composites, Journal of Asian Ceramic Societies, Vol. 1, pp. 30-34 (DOI: 10.1016/j.jascer.2013.01.002).
22. A.Sá Ribeiro Ruy, G.Sá Ribeiro Marilene, and K. Sankar, Waltraud M. (2016), Kriven Geopolymer-Bamboo Composite-A Novel Sustainable Construction Material, Construction and Building Materials, Vol. 123, pp. 501-507 (DOI: 10.1016/j.conbuildmat.2016.07.037).
23. V. Agopyan, H. Savastano Jr, V.M. John, and M.A. Cincotto, (2005), Developments on Vegetable Fibre-Cement based Materials in Sago Paulo, Brazil: An Overview, Cement and Concrete Composites, Vol. 27, pp. 527-536.
24. R.A.J. Malenab, J.P.S. Ngo, and M.A.B. Promentilla, (2017), Chemical Treatment of Waste Abaca for Natural Fiber-Reinforced Geopolymer Composite, Materials, Vol. 10, pp. 579-xxx (DOI:10.3390/ma10060579).
25. J. He, Y. Jie, J. Zhang, Y. Yu, and G. Zhang, (2013), Synthesis and Characterization of Red Mud and Rice Husk Ash-based Geopolymer Composites, Cement and Concrete Composites, Vol. 37, pp. 108-118.
26 S. Detphan and P. Chindaprasirt, (2009), Preparation of Fly Ash and Rice Husk Ash Geopolymer, International Journal of Minerals Metallurgy and Materials, Vol. 16, pp. 720-726 (DOI: 10.1016/S1674-4799(10)60019-2).
27 M. Younes, H. Abdel-Rahman, and M.M. Khattab, (2018), Utilization of Rice Husk Ash and Waste Glass in the Production of Ternary Blended Cement Mortar Composites, J. Build. Eng., Vol. 20, pp. 42-50.
28. E. Rill, D.R. Lowry, and W.M. Kriven, (2010), Properties of Basalt Fiber Reinforced Geopolymer Composites, In book: Strategic Materials and Computational Design: Ceramic Engineering and Science Proceedings, Vol. 31, (DOI: 10.1002/9780470944103.ch6),

29. M.M. Camargo, E.A. Taye, J.A. Roether, D.T. Redda and A.R. Boccaccini, (2020), A Review on Natural Fiber-Reinforced Geopolymer and Cement-Based Composites, Materials, Vol. 13(20), pp. 4603-xxx (DOI:10.3390/ma13204603).
30. R Zulfiati, Saloma, and Y Idris, (2019), Mechanical Properties of Fly Ash-Based Geopolymer with Natural Fiber, (SENTEN 2018 - Symposium of Emerging Nuclear Technology and Engineering Novelty), IOP Conf. Series: Journal of Physics: Conf. Series, Vol. 1198, pp. 082021-xxx (DOI:10.1088/1742-6596/1198/8/082021).
31. D. Wattanasiriwech, T. Munmueangkham, and S. Wattanasiriwech, (2018), Impact Strength and Physical Properties of Geopolymer Composites Reinforced with Bagasse Cellulose Fibers, The 2018 World Congress on Advances in Civil, Environmental, and Materials Research (ACEM18), Songdo Convensia, Incheon, Korea.
32. Siti Noorbaini Sarmin, Lightweight geopolymer wood composite synthesized from alkali-activated fly ash and metakaolin, November 2016, Jurnal Teknologi 78(11):11-2016, DOI: 10.11113/.v78.8734,
33. S. Elavarasan and K. Dhanadeepa, (2017), Experimental Investigation on Hybrid Fiber Reinforced Geopolymer Concrete, International Journal of Innovative Research in Science, Engineering and Technology, Vol. 6, pp. 4114-4118 (DOI:10.15680/IJIRSET.2017.0603192).
34. M. Ohno and V. Li, (2014), A Feasibility Study of Strain Hardening Fiber Reinforced Fly Ash-based Geopolymer Composites, Construction and Building Materials, Vol. 57, pp. 163-168. (DOI: 10.1016/j.conbuildmat.2014.02.005).
35. B.V. Rangan, D. Hardjito, S.E. Wallah, and D.M.J. Sumajouw, (xxx), Studies on Fly Ash-based Geopolymer Concrete, Geopolymer: Green Chemistry and Sustainable Development Solutions, Journal Details, pp. 133-138.
36. T. Subramani and P. Sakthivel, (2016), Experimental Investigation on Flyash based Geopolymer Bricks, International Journal of Application or Innovation in Engineering & Management (IJAIEM), Vol. 5, pp. 216-227.,
37. F. Shaikh and S. Haque, (2018), Behaviour of Carbon and Basalt Fibres Reinforced Fly Ash Geopolymer at Elevated Temperatures, International Journal of Concrete Structures and Materials, Vol.xxx, pp. 1-12.
38. M. Panigrahi, S. Mohanty, R.I. Ganguly, and R.R. Dash, (2018), An Advanced Cured High Carbon Ferrochrome Slag (HCFCS) Geopolymer (GP): A Constructional Materials, IOP Conf. Series: Materials Science and Engineering, Vol. 410, pp.012002-xxx (doi:10.1088/1757-899X/410/1/012002).
39. M. Panigrahi, S. Mohanty, R.R.Dash, and R.I. Ganguly, (2008), Development of Novel Constructional Material from Industrial Solid Waste as Geopolymer, IOP Conf. Series: Materials Science and Engineering, Vol. 410, pp.012012-12 (doi:10.1088/1757-899X/410/1/012012).

40. D.P. Mishra and S.K. Das, (2010), A Study of Physico-Chemical and Mineralogical Properties of Talcher Coal Fly Ash for Stowing in Underground Coal Mines, Mater. Characterization, Vol. 61, pp. 1252-xxx.
41. G. Naidu, P.B.J. Kumar, S. Adisesh, and P.V.V. Satyanarayana, (2012), A Study on Strength Properties of Roller Compacted Concrete with 30% Replacement of OPC 53 Grade Cement by Flyash, Int. J. Comput. Engg. Research, Vol. 2, pp. 912-xxx.
42. R. Vora, D. Prakash, and V. Urmil, (2013), Parametric Studies on Compressive Strength of Geopolymer Concrete, Procedia Engg, Vol. 51, pp. 210.
43 A.S. Ahmed, Md. S. Islam, A. Hassan, M.K.M. Haafiz, Kh. N. Islam, and R. Arjmandi, (2014), Impact of Succinic Anhydride on the Properties of Jute Fiber/Polypropylene Biocomposites, Fibers and Polymers, Vol.15, pp. 307-314 (DOI: 10.1007/s12221-014-0307-8).
44 Y. Seki, (2009), Innovative Multifunctional Siloxane Treatment of Jute Fiber Surface and Its Effect on the Mechanical Properties of Jute/Thermoset Composites, Materials Science and Engineering: A, Vol. 508, pp.247-252 (https://doi.org/10.1016/j.msea.2009.01.043).
45. Anu Jose, M.R. Nivitha, J. Murali Krishnan, and R.G. Robinson, (2020), Characterization of Cement Stabilized Pond Ashusing FTIR Spectroscopy, Construction and Building Materials, Vol. 263, pp. 120136-xxx (https://doi.org/10.1016/j.conbuildmat.2020.120136).
46. Zuhua Zhang, Hao Wang, and John L. Provis, (2012), Quantitative study of the reactivity of fly ash in Geopolymerization by FTIR, Vol. xxx, pp. 154-166 (https://doi.org/10.1080/21650373.2012.752620).
47. A. Atmakuri, A. Palevicius, L. Kolli, A. Vilkauskas, and G. Janusas, (2021), Development and Analysis of Mechanical Properties of Caryota and Sisal Natural Fibers Reinforced Epoxy Hybrid Composites, Polymers, Vol.13, pp. 864-19. (https://doi.org/10.3390/polym13060864).
48. Saravana, K. Raja Mohan, and A. Sumathi, (2017), Effect of Fly Ash in Fiber Reinforced Concrete Composites, Jordan Journal of Civil Engineering, Vol. 11, pp. 30-39.
49. D.K. Rajak, D.D. Pagar, P.L. Menezes and E. Linul, (2019), Review Fiber-Reinforced Polymer Composites: Manufacturing, Properties, and Applications, Polymers, Vol. 11, pp. 1667-37; (doi:10.3390/polym11101667).
50. A.J. Reffin, M. Karuppasamy, S. Manikandakumar, and G. Baskar Singh, (2021), An Experimental Study of Jute Fiber and Partially Added Fly Ash with Ordinary Portland Cement in Concrete, International Research Journal of Engineering and Technology (IRJET), Vol. 08, pp. 1435-1439.
51. M.M. Kabir, Md M. Islam, and H. Wang, (2013), Mechanical and Thermal Properties of Jute Fibre Reinforced Composites, Journal of multifunctional materials, Vol. xxx, pp.71-77, (DOI:10.12783/issn.2168-4286/1.1/Islam)
52. M. Karahan, F. Ozkan, K. Yildirim, and N. Karahan, (2016), Investigation of the Properties of Natural Fibre Woven Fabrics as a Reinforcement Materials

for Green Composites, Fibres & Textiles in Eastern Europe, Vol. 24, pp. 98-104 (DOI: 10.5604/12303666.1201138).
53. E. Jayamani, Md. R Rahaman, D.A. Benhur, M.K. Bin Bakari, A Kakar, and A Khan, (2020), Comparative Study of Fly Ash/Sugarcane Fiber Reinforced Polymer Composite Properties, BioResources, Vol. 15, pp. 5514-5531.
54. S. K. Acharaya, P. Mishra, and S.C. Mishra, (2008), Effect of Environment on the Mechanical Properties of Fly Ash-Jute-Polymer Composite, Indian Journal of Engineering and Materials Science, Vol. 15, pp. 483-488.

7

Corrosion of Pond Ash (PA)-Based Geopolymer Products

Slipika Panda[1], Muktikanta Panigrahi[2*], Ratan Indu Ganguly[3] and Radha Raman Dash[4]

[1]*Department of Civil Engineering, Gandhi Institute of Engineering and Technology University, Gunupur, Odisha, India*
[2]*PG Department of Materials Science, Maharaja Sriram Chandra Bhanja Deo University, Keonjhar Campus, Odisha, India*
[3]*Department of Metallurgical Engineering, National Institute of Technology, Raurkela, Odisha, India*
[4]*CSIR-National Metallurgical Laboratory, Jamshedpur, Jharkhand, India*

Abstract

The corrosion behavior of PA-based geopolymer mortar and concrete under different conditions (saline water, HCl solution, H_2SO_4 solution, H_3PO_4 solution, normal water, distilled water) is discussed in this chapter. Strength, morphology and thermal properties were compared for exposed and unexposed GP motor and concrete. The strengths obtained were 24.1 MPa and 29.9 MPa for PA-based GP mortar and PA-based GP concrete, respectively, after 75 days exposure to saline water. Similarly, the strengths obtained were 24.3 MPa and 30.1 MPa for PA-based GP mortar and PA-based GP concrete after 75 days exposure to HCl media water. A similar corrosion test is performed in distilled water media. The strength was found to be 24.1 MPa (PA-based GP mortar) and 30.0 MPa (PA-based GP concrete) after 75 days corrosion in distilled water. Thus, there is no difference in strength properties observed in those two materials after exposure to different media. The strength before corrosion in water was found to be 24 MPa (PA-based GP mortar) and 29.7 MPa (PA-based GP concrete). Results indicate that there is no change in strength. There is no change in density results for exposed and non-exposed samples corroborated by strength results. SEM images and DSC isotherms of GP products indicated the occurrence of condensation polymer reaction

**Corresponding author*: muktikanta2@gmail.com

Muktikanta Panigrahi, Ratan Indu Ganguly and Radha Raman Dash (eds.) Development of Geopolymer from Pond Ash-Thermal Power Plant Waste: Novel Constructional Materials for Civil Engineers, (195–226) © 2023 Scrivener Publishing LLC

during curing. TGA analyses did not show any difference in behavior between the materials (before and after exposure of PA-based GP mortar and concrete).

In conclusion, it may be said that the strength values for these two materials are comparable to that of standard M15 grade mortar used in construction.

***Keywords*:** OPC, pond ash, geopolymer, concrete, corrosive media, AAS, SEM, corrosion

7.1 Introduction

Up to the 20th century, ordinary Portland cement (OPC) concrete was used as building material worldwide, producing 2.8 million tons of CO_2 annually. Studies predict that by 2050 OPC production will sharply rise to 5.5 gigatons per year [1–3], which will have a significant impact on the environment due to massive consumption of natural resources [1, 2, 4]. So, it can be said that cement production alone can contribute approximately 7% of human-driven global CO_2 emissions [1, 4–8].

The global warming problem and greenhouse gas effect has led to numerous investigations in an attempt to develop new binding or cementitious construction materials for reducing CO_2 emissions during the production of OPC, giving rise to the development of new alternative materials [8]. These newly developed alternative materials, which were first developed by Joseph Davidovits in 1982 [9], are called geopolymers. Geopolymers are tetrahedral three-dimensional inorganic aluminum silicate inorganic polymers composed of [AlO_4] and [SiO_4], which are mainly prepared from aluminum silicates or industrial wastes such as pond ash [10, 11], fly ash [3, 12–14], natural pozzolana [15, 16], metakaolin [17, 18], natural zeolites [4, 19–21], etc. Geopolymers are prepared by the activation of aluminum silicates in an alkali-silicate solution [9, 22, 23]. Geopolymers are a class of aluminosilicate binding materials, which are synthesized by a condensation polymerization process called geopolymerization of aluminosilicate minerals (such as industrial solid waste, calcinated clays, natural minerals and so on) with solutions of alkaline activator [24]. In recent years, they gradually started being considered as potentially revolutionary materials in order to obtain the great advantages presented by their thermal, mechanical and chemical resistance properties among others [17].

Corrosion of hydrated cement-based materials, i.e., structural materials, undergo a number of complex processes which reduce engineering properties such as strength, elastic modulus and durability. Such complex processes may happen due to various physical, chemical, and physicochemical factors and their interactions [25–33], reflecting the peculiarities of both material aspects and external factors (service conditions).

A simple description of the physicochemical processes that take place during corrosion of porous hydrated inorganic materials is relatively difficult and usually consists of three steps. The first step is the convective and diffusional transport of aggressive medium (gas or liquid) through interconnected pores to the corrosion front. In the second step, the aggressive substance reacts with the material, resulting in the formation of some soluble and/or insoluble products. Deposition of insoluble products in the corroded parts and/or transport of soluble products through corroded layer to the aggressive medium can be considered the third step. The process of corrosion is complicated for the following reasons: Accurate characterization of the pore system in the process and difficulties due to slow diffusion of ions and rate of permeation of potentially deleterious substances across a porous cementitious matrix are intuitively related to the volume of pores, pore size and shape and pore interconnectivity (open and closed pores). Pore radii in cement pastes span a huge range from a few nanometers to nearly 100 microns (gel and capillary pores). In more complex, composite systems, such as concrete, a pore size distribution of entrained and entrapped air voids is also present [27–33]. Such a complex and irregular structure in concrete causes the stochastic character of concentration patterns and flows [27–33].

In cement mortar and concrete, the reaction of aggressive molecules and/or ions with reactive aggregate and the phenomena taking place at the interface between cement hydrates and aggregate make the problem more complicated. The most common causes of corrosion of concrete structures include salt attack (mostly sulfate and seawater), acid attack, including carbonation, alkali-silica reaction, efflorescence, and freeze-thaw/frost damage. There are some other causes which occur in industrial environments. These include high temperature attack, abrasion (cavitation and erosion), and attack by specific chemical or biological agents. A quick review of the literature in this field (durability of cement-based materials) shows that much effort has been spent by researchers during the past fifty years. For instance, Struble [27–33] cited 638 references drawn from three conference proceedings and various journal articles only for the year 1987 alone. The related literature has been presented in a number of different information sources. It should be noted that despite such a voluminous amount of literature, comparing results and acquiring relevant information is usually difficult. This is firstly because of differences in experimental conditions, including temperature, concentration and composition, static and dynamic conditions, time period of experiment, etc.; and secondly, because, in practice, corrosion of concrete is usually due to two or more causes working simultaneously.

In spite of the importance of the phenomenon, there are relatively few publications available on this specific subject in the literature compared to the other common types of corrosion of cement-based materials. In addition, most of the existing publications also deal with the case of attack by carbonic acid (i.e., carbonation or aggressive effects of CO_2). On the other hand, development of new cements with improved properties necessitates detailed investigations on their durability in aggressive environments [25–33].

Many researchers have studied the corrosion behavior of OPC concrete, mainly for corrosion of steel reinforcement embedded in concrete. This is important for steel-reinforced concrete structure near coastal regions where the climate is characterized by contaminants such as chlorine ions (Cl^-) and carbon dioxide (CO_2) [34–39]. Usually, embedded steel gets detached from the main structure and, consequently, destroys the structure. The use of corrosion inhibitors is a method to prevent and delay the appearance of corrosion in concrete-reinforced bars [40, 41]. One of the possible definitions of corrosion inhibitors is given by ISO 8044: 1989, which establishes that the material is a chemical substance, in certain quantity, that slows down the corrosion rate without any change in the concentration of any other corrosion agent [42]. Recently, the corrosion inhibitors have become easy to implement for the protection of reinforced steel in concrete but are to be considered a variable. These additives must not only prevent or delay the appearance of the corrosion, but also must not have any harmful effect on the properties of concrete [38, 43].

Dewi and Ekaputri [44] presented some results of an experimental study on bond strength of plain bar embedded in geopolymer concrete. Nine cubic specimens (150 × 150 × 150 mm) were prepared to measure bond strength and slip between reinforcement and concrete.

Al Bakri *et al.* [45] evaluated the mechanical and physical properties of GP prepared by them. They observed high strength, low shrinkage, good thermal and chemical resistance in their materials.

Farhana *et al.* [46] described the corrosion behavior of reinforced bar coated with GP paste. The studies were done with and without curing of GP paste. They used tap water as media and potential in open-circuit potential (OCP) method for assessing extent of corrosion. The reinforcement steel bar in cured geopolymer showed more positive potential values as compared with uncured samples. It was noted that for samples soaked in tap water, the potential values fluctuated. From the Pourbaix diagram, all the observed potential values were found to be located in the passivity region. The oxide formed provided a passive layer to protect the reinforcement bar from corrosion.

Provisa *et al.* [47] described the development process for alkali-activated materials (AAMs). They evaluated long-term performance of steel-reinforced GP structures. The durability of AAMs for such applications strongly depended on the corrosion behavior of the embedded steel reinforcement. The experimental data in the literature are limited and, in some cases, inconsistent.

Allahverdi and Škvára [48] investigated the mechanism of corrosion of hardened paste of geopolymer cements at relatively high concentrations of sulfuric acid (pH ≈ 1). The process of corrosion occurred in two subsequent steps. The first step involves the ion exchange reaction between the charge compensating cations of the framework and electrolytic solution. The electrophilic attack of acid protons results in the ejection of tetrahedral aluminum from the aluminosilicate framework. In the subsequent step, exchange calcium ions diffuse toward the acid solution and react with counter-diffusing sulfate anions. This results in the formation and deposition of gypsum crystals inside corroding layer. Deposition of gypsum crystals inside the corroding matrix provides a protective measure inhibiting the total process of deterioration.

Manickavasagam and Mohankumar [49] developed an environmentally friendly green geopolymer concrete. They observed improved compressive strength with materials designated as GM20 and GM30. They also investigated various properties, such as durability, water absorption, acid resistance and corrosion resistance, with constant voltage polarization test. The result obtained from high-calcium fly ash-based GPC was found to be superior in durability properties when compared with OPC-based concrete.

Asmara [50] focused on an experimental investigation to identify the effects of fly ash on the electrochemical process of corrosion in concrete structure during the curing time. The extent of corrosion was measured with the help of a potentiostat. Measured rest potential, polarization diagram, and corrosion rate were determined. They also studied the effect of variation of cement and fly ash ratio on corrosion data. After being cured for 24 h and 65 °C, samples were immersed in a 3.5% solution of NaCl for 365 days for electrochemical measurement. The results showed good corrosion resistance of their materials.

Udhaya Kumar and Jayadurgalakshmi [51] assessed the technical viability of the use of geopolymer by conducting a corrosion test on a steel reinforcement bar embedded in geopolymer material. The available alkalinity of geopolymer material initially was suspected to be harmful for alkali-silica reaction, but it was found to be beneficial for maintaining passivity of the steel bar in concrete. The current density was applied to accelerate corrosion influences on bond strength.

Aulia *et al.* [52] studied corrosion resistance of high-strength reinforced concrete under different environmental conditions. They measured the potential level and corrosion rate of their materials. Later, they studied the same structural materials by proportionate replacement of cement with the fly ash-based geopolymer. Corrosion potential measurement was carried out with the half-cell potential mapping method; and corrosion rate was assessed by linear polarization resistance method. They used different corrosion media, i.e., 3.5% NaCl (artificial seawater) and shallow well water. Corrosion rate for a few composite structures, i.e., PFA, CFA, POSFA, and HSCWFA, etc., were found to be the lowest.

Rambabu *et al.* [53] studied corrosion behavior of geopolymer in sulphate and acid media. They used open-circuit potential (OCP) method for corrosion measurement. They also evaluated the tensile behavior of corroded structure at different stages of corrosion of their materials.

Bayuaji *et al.* [54] studied the effect of an alkali solution environment on geopolymer concrete. They observed durability of GP concrete structure in corrosive seawater environment. Tensile strength of corroded GP concrete was also measured by them.

Bhardwaj *et al.* [55] coated mild steel substrate with advanced phosphatic geopolymer using the spray coating technique. The coated plates were tested for adhesion strength, water resistance, fire protection and corrosion resistance. Results indicated that coating of mild steel substrate in two passes (with thickness 100 ± 15 μm) showed better adhesion and promising corrosion resistance for mild steel substrate under seawater conditions.

Yodsudjai [56] investigated corrosion behavior of reinforcement for several concrete compositions. Different cement types and supplementary cementitious materials (SCMs) were used for their concrete structure. Corrosion properties of the structure were evaluated by adopting different methods, *viz.*, half-cell potential, corrosion current density, electrical conductivity topography and reinforcement cross-sectional loss. Results showed that concrete with SCMs possessed the highest corrosion resistance, followed by concretes with different cement types and concretes with special admixtures, respectively.

Sikora *et al.* [57] investigated a chemical attack on suitably coated geopolymer mortar/concrete in corrosive environment (10% inorganic/organic acids and saturated solution of sodium chloride). They measured weight loss and compressive strength of corroded geopolymer composite and concrete. A coated concrete sample showed considerable weight loss as well as compressive strength.

Shaikh [58] described chloride-induced corrosion durability of geopolymer concrete reinforced with steel. They use geopolymer material prepared with different ratios of sodium silicate (Na_2SiO_3) and NaOH solutions. Specimens underwent cyclic treatments (wetting and drying regime) for two months. In the wet cycle, specimens were immersed in NaCl salt solution (3.5% by weight) for 4 days. In the dry cycle, specimens were placed in open air for three days. For the measurement of corrosion rate, potential values were recoded using copper/copper sulphate ($Cu/CuSO_4$) half-cell (ASTM C-876). Conclusions drawn from their results indicate that the corrosion resistance of geopolymer concretes is superior to OPC concrete.

Lakshmi Narayanan and Vasugi [59] investigated the corrosion rate for ground granulated blast furnace slag (GGBS) geopolymer which is used in reinforced concrete (RC) building materials. They used different methods for corrosion rate measurement. For RC building materials, steel plays an important role because its strength and durability properties decide the stability of the structures. Chloride ion present in the concrete forms a passive layer over the embedded steel in concrete, which prevents corrosion for the structure. Thus, they concluded that geopolymer concrete is a new alternative to conventional concrete.

Udhaya Kumar and Jayadurgalakshmi [60] investigated corrosion in M30 grade geopolymer concrete. Corrosion of reinforcement has been described as a predominant factor causing widespread premature deterioration of concrete in coastal marine environment. Corrosion of steel reinforcement bar embedded in geopolymer matrix was used for their confirmation of technical viability. After 28 days of ambient curing, RCC and GPC beams were subjected to 0%, 5%, 7.5% and 10% corrosion by accelerated corrosion test. Corrosion current rate varied from $1mA/cm^2$ to $4mA/cm^2$. Ultrasonic pulse velocity test was conducted on RCC and reinforced GPC beams. Flexural behavior of corrosion-induced RCC beams and reinforced GPC beams was studied.

Wu *et al.* [61] produced geopolymer concrete with increased resistance to acid attack. They prepared geopolymer concrete using aluminosilicate materials, metakaolin pozzolana, sodium hydroxide, and alkali activator. Durability performance of geopolymer cement in acid media (1.0 M sulfuric acid) was evaluated after exposure in the media for 7/28/90 days. Results showed that the presence of nanosilica and nanoclay in raw materials improved performance of geopolymer concrete.

Villalba *et al.* [62] explored the possibility of using natural zeolite-based geopolymer. They evaluated thermal stability and mechanical properties of these materials. Natural zeolite was geopolymerized by treating with alkali (8 M NaOH solution), 3 wt% calcium hydroxide $Ca(OH)_2$ and binder. Green materials were subsequently cured for 24 h at 40 °C followed by curing at room temperature for 6 days. The prepared materials were then characterized with XRD, SEM-EDS, etc. They also studied the effect of adding a migratory corrosion inhibitor (MCI-2005 NS) on mechanical properties of geopolymers.

Ariza-Figueroa *et al.* [63] prepared ternary ecological concrete (TEC). In their structure, they replaced ordinary Portland cement (OPC) with sugar cane bagasse ash (SCBA) (10%, 20%, 30%) and silica fume (SF). Concrete structures were reinforced with AISI 304 stainless steel and AISI 1018 carbon steel rebars. An electrochemical corrosion study was carried out on the structure by corrosion potential (Ecorr) method. They observed that TEC mixture made with a 20% blend of sugar cane bagasse ash-silica fume (SCBA-SF) to steel-reinforced OPC gave the best corrosion performance.

Mohammed *et al.* [64] developed geopolymer cement using high-calcium fly ash (HCFA). They activated HCFA with anhydrous sodium metasilicate powder. Green geopolymer cement was cured under ambient conditions. Density, flowability, setting time, compressive strength, splitting tensile strength and molar ratio impact were evaluated for geopolymer developed. Structural assessment of geopolymer was done by FESEM, FTIR and XRD.

An extensive literature survey was included in the previous chapter. It is evident from the survey that geopolymer can play an important role in replacement of conventional cement used for common structures. Geopolymer concrete will be used in buildings in the near future. Different authors have shown that geopolymer can be prepared from industrial waste materials such as fly ash, pond ash, bottom ash, slags, etc. Structures built with geopolymer are subjected to various environments for their corrosion test. It is found that geopolymer structures retain their strength properties and durability in coastal environment. Microstructure aspects have proven structure stability in corrosive atmosphere.

In the present work, geopolymer was prepared from pond ash. It was corrosion tested in order to reach the final stage of development and production of the materials. The authors mostly concentrated on corrosion aspects of the materials-in acid media, saline water, normal water and distilled water medium. The properties of structure built with geopolymer concrete and mortar are discussed along with the different characterization methods adopted to understand the strengthening mechanism in the materials.

7.2 Experimental Details

7.2.1 Chemicals and Materials

Pond ash (PA), the main ingredient of geopolymer, was collected from NALCO Damanjodi, Odisha [11]. The sand and aggregates used were collected from a local market. Sodium hydroxide (NaOH) obtained from M/S Loba Chemicals, India, was used as an activator. Sodium silicate (Na_2SiO_3) used for the purpose of polymerization was procured from Merck, India. Water-soluble plasticizer (Sika) was purchased from a reputable firm located in the southern part of India (Visakhapatnam market). Plasticizer usually enhances plasticity of the product, i.e., geopolymer/geopolymer mortar/concrete. Details of the required materials required for this synthesis are described in the following section.

7.2.2 Preparation of Pond Ash-Based Geopolymer Products

7.2.2.1 Pond Ash-Based Geopolymer Mortar Preparation

Preparation of pond ash-based geopolymer mortar was achieved by chemical approach through solid-state method. In this process, GP made from PA was mixed with sand in appropriate quantity to prepare GP motor [11]. For proper mixing, it was carried out in a mixer. All the consecutive steps followed for preparation of GP mortar are sequentially discussed. The steps include grinding, mixing, ramming and curing (using hot air oven) [11].

Step 1: Raw Material Preparation
As-received pond ash is dried at 120 °C in hot oven. Then it is ground and sieved through 240 meshes. Collected sand is sieved and dried at 120 °C in hot oven. Processed pond ash and sand are kept separately. Sodium silicate and sodium hydroxide are combined to use as geopolymer activator. Water-soluble plasticizer is used to increase plasticity.
Step 2: Geopolymer Mortar Preparation
The sieved pond ash is mixed with NaOH, Na_2SiO_3, a small quantity of water (6%) and water-soluble plasticizers (Sika); and a gel-like substance is formed. The distinct compositions of the mixtures are further mixed with the requisite amount of sand to form mortar. Proportions of mortar compositions are shown in Table 7.1.
Step 3: Molding, Casting and Compaction
As-prepared gel-like materials are put in REMI system. The REMI mold is made from iron in cylindrical form, with 50 mm diameter and 70 mm

Table 7.1 Compositions of PA-based mortar mixtures (Mortar = 1:3).

Sample code	PA (21.125%) + sand (63.375%)	Sod. silicate (SS)	Alkali (8 M, NaOH) (SH)	Water-soluble plasticizer (Sika)
S1	85%	12%	3%	1–2 ml
S2	85%	12%	3%	1–2 ml
S3	85%	12%	3%	1–2 ml

Figure 7.1 Photograph of (a) PA-based geopolymer mortar, and (b) Concrete samples for corrosion study.

height. Samples are rammed 20 times. They are then left for 3 min before being demolded. These are known as "green" samples.

Step 4: Curing Process

The cast samples are cured at 70 °C for 24 h in an oven. The cured samples are sealed in plastic zipper bags to prevent moisture loss and then stored until the samples are tested. The PA-based GP mortar (a) and concrete (b), shown in Figure 7.1, are then ready for the corrosion study.

7.2.2.2 Pond Ash-Based Geopolymer Concrete Preparation

The concrete preparation process was achieved by a chemical approach through solid-state method. Both sand and aggregates were mixed in the mixer at the time of mixing with PA-based geopolymer in proper proportions. The consecutive steps of concrete preparation are grinding, mixing, ramming and curing (in hot air oven) [11].

Step 1: Raw Material Preparation
Pond ash is dried at 120 °C in hot oven. Then it is ground and sieved through 240 meshes. Collected sand is sieved and dried at 120 °C in hot oven. As-purchased aggregates are dried at 120 °C. Components are kept separately. Sodium silicate and sodium hydroxide are combined to use as geopolymer activator. Water-soluble plasticizers are used to increase plasticity.
Step 2: Geopolymer Concrete Preparation
Processed pond ash is mixed with NaOH, Na_2SiO_3, a small quantity of water (6%) and water-soluble plasticizers (Sika); and a gel-like substance is formed. These are further mixed with the requisite amount of aggregates and sand to form concrete. Different proportions of concrete are shown in Table 7.2.
Step 3: Molding, Casting and Compaction
As-prepared gel-like materials are put in REMI system. The REMI mold is made from iron in cylindrical form, with 50 mm diameter and 70 mm height. The samples are rammed 20 times. They are then left for 3 min before being demolded. These are known as "green" samples.
Step 4: Curing Process
The cast samples are cured at 70 °C for 24 h in an oven. The cured samples are then sealed in plastic zipper bags to prevent moisture loss and then stored until the samples are tested. PA-based GP concrete is shown in Figure 7.1b, which is ready for corrosion and other studies.

Table 7.2 Compositions of PA-based concrete mixtures (Concrete = 1.0:1.5:3.0).

Sample code (Three Sets)	PA (15.36364%) + sand (23.04546%) + aggregates (46.09092%)	Sod. silicate (SS)	Alkali (8 M, NaOH) (SH)	Water-soluble plasticizer (Sika)
S1	85%	12%	3%	1–2 ml
S2	85%	12%	3%	1–2 ml
S3	85%	12%	3%	1–2 ml

7.2.3 Characterizations of Pa-Based Geopolymer GP Mortar/Concrete (Before and After) Corrosion

The extent of corrosion is marked as loss in weight of GP products after removal of samples from corrosive media. The GP products are initially weighed and then they are dipped into containers (5-liter capacity) filled with different corrosive media (1.0 M HCl/1.0 M H_2SO_4/1.0 M H_3PO_4/3.5% NaCl solution/normal water/distilled water). Samples were exposed for different lengths of time. After a regular interval, samples were removed in different batches and were then dried by exposing in a heating oven at 120 °C. Then, the weight of dried samples was measured. Bulk densities of samples were estimated using the expression $\rho = M/V$, where M = mass and V = volume of GP materials. The composition of liquid molecules was checked on a regular basis by flame atomic absorption spectrometry (F-AAS). Deuterium background correction was utilized. A Z300 Benchtop Centrifuge (Hermle LaborTechnik, Germany) was employed for centrifuging sample solution in MIBK extraction. Gold specks obtained by cyanidation process were weighed using a UMT2 Micro Balance (Mettler-Toledo, Switzerland). The instrument parameters were as follows: lamp type = hollow cathode lamp, flame type = air acetylene, background correction = deuterium, wave length = 242.8 nm, slit width = 1.0 nm, lamp current = 4 mA, measurement time = 1.5 sec, air flow = 3.51/min, and acetylene flow = 1.51/min. In this study, compressive strength tests were performed on 70 × 50 mm cylindrical specimens using a digital operated 2000 KN Compression Testing Machine (Mechatronics Control System; Model: MECH-CS/UTE 20T). A set of three specimens for each sample were tested after curing for 1 day. Experiments were replicated a few times to ascertain the validity of our test results. Compressive strengths were measured at higher temperatures (70 °C, 80 °C, 90 °C). Microstructural characterization and *in-situ* elemental analyses were carried out by SEM/EDX and a Carl Zeiss Supra-40 Scanning Electron Microscope. This helped to understand the morphology of the geopolymer. Fourier transform infra-red (FTIR) spectroscopy analyses were carried out on all the samples; and the FTIR absorption spectra were recorded for all the samples in the ranges from 4000–400 cm^{-1} at room temperature. A sample of 4.0 mg was mixed with 200-mg KBr in an agate mortar and then was pressed into pellets of 13-mm diameter. For each sample, the FTIR spectrum is normalized with the blank KBr pellet.

The enthalpy of fusion of as-prepared PA-based GP mortar/concrete at 70 °C curing temperature and 24 h curing time was calculated from differential scanning calorimetric (DSC) experiment. The parameters such as heating rate = 10 °C/min, temperature range = 50 to 300 °C, and nitrogen environment (N_2) were set. The thermal stability of the as-prepared PA-based GP mortar/concrete at 24 h curing and 70 °C curing temperature was determined by thermogravimetric analysis using a Perkin Elmer Pyris Diamond analyzer at a heating rate of 10 °C/min in nitrogen environment from 50–800 °C. Both thermal experiments (DSC and TGA) were performed using a Perkin Elmer Pyris Diamond analyzer.

7.2.4 Results and Discussion

For the corrosion study, two parameters were chosen to show any differences in physical properties that occurred after exposing the materials for different lengths of time in different corrosive media. Change in weight and bulk density can be correlated with the stability and integrity of the structure. The present materials were proposed for use as structural material replacing conventional Portland cements.

Geopolymer mortar samples and geopolymer concrete samples were prepared in batches using similar treatment combinations. From each batch, representative samples were chosen for corrosion testing.

Figures 7.2a and 7.2b shows bar charts of unexposed and exposed (with different length of exposing time) PA-based GP mortar sample(s) in distilled water medium. Similar bar charts are plotted for GP concrete (Figures 7.2c & 7.2d) after being exposed in distilled water media. It is evident that there is not much change in weight and bulk density of exposed materials if compared with unexposed materials. Thus, the materials are found to be resistant to distilled water.

Figures 7.3 shows the comparison in weight change and bulk density between unexposed and exposed PA-based GP Mortar/Concrete samples in normal water media. Figures 7.3a and 7.3b show bar charts of unexposed and exposed PA-based GP Mortar sample(s) in normal water media in different lengths of exposure time. Similar bar charts are plotted for GP concrete (Figures 7.3c & 7.3d) after being exposed in normal water media. It is evident that there is no significant change in weight and bulk density of exposed materials if compared with unexposed materials. Thus, the materials are found to be resistant to normal water.

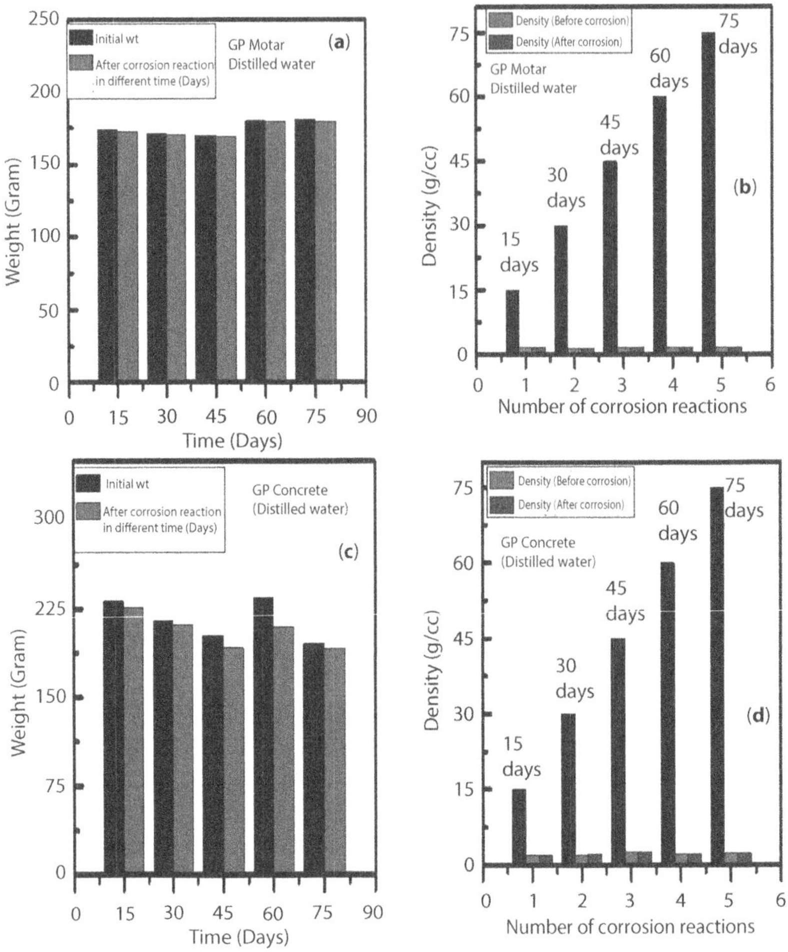

Figure 7.2 Comparison of weight change and bulk density between unexposed and exposed PA-based GP Mortar Concrete samples in distilled water medium.

Figures 7.4a and 7.4b shows bar charts of unexposed and exposed (with different lengths of exposing time) PA-based GP Mortar sample(s) in saline water media. Similar bar charts are plotted for GP concrete (Figures 7.4c & 7.4d) after being exposed in saline water media. It is evident that there is not much change in weight and bulk density of exposed materials if compared with unexposed materials. Thus, the materials are found to be resistant to saline water.

Figures 7.5a and 7.5b shows bar charts of unexposed and exposed (for different lengths of exposure time) PA-based GP Mortar sample(s) in

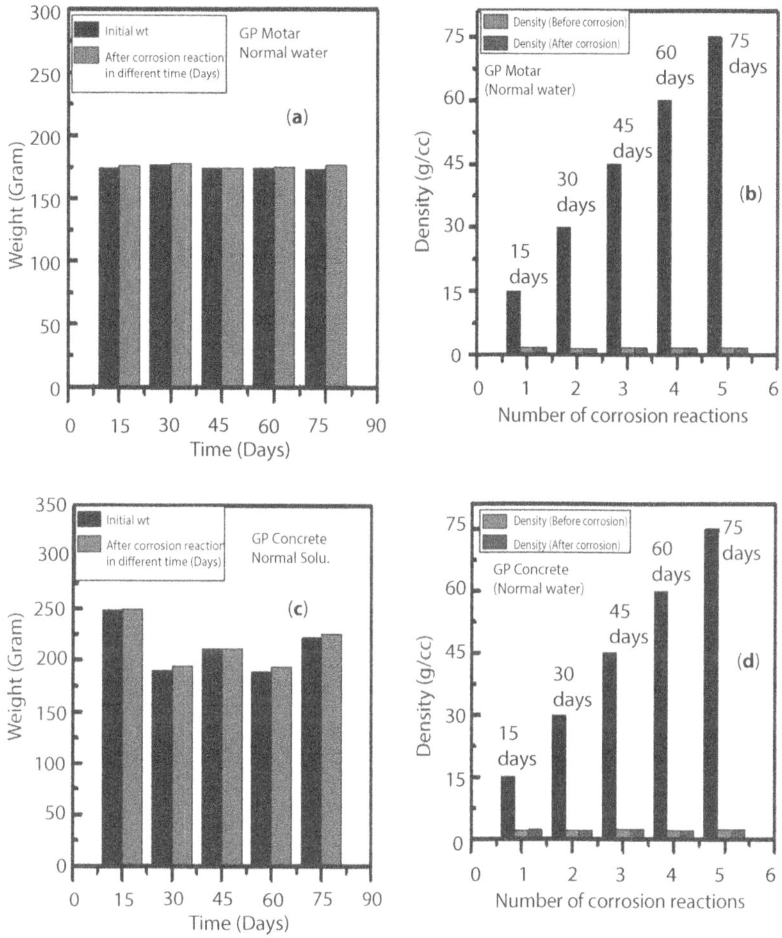

Figure 7.3 Comparison in weight change and bulk density between unexposed and exposed PA-based GP Mortar/Concrete samples in normal water media.

H_2SO_4 acid media. For GP concrete (Figures 7.5c & 7.5d), identical bar charts are plotted after being exposed in H_2SO_4 acid media. Not much change in weight and bulk density is observed in exposed materials in comparison to unexposed materials. Thus, the materials are found to be resistant to H_2SO_4 acid media.

Figures 7.6a and 7.6b shows bar charts of unexposed and exposed (for different lengths of exposure time) PA-based GP Mortar sample(s) in H_3PO_4 acid media. Similar plots of bar charts for GP concrete (exposed in H_3PO_4 acid media) is shown in Figures 7.6c and 7.6d. There is not much

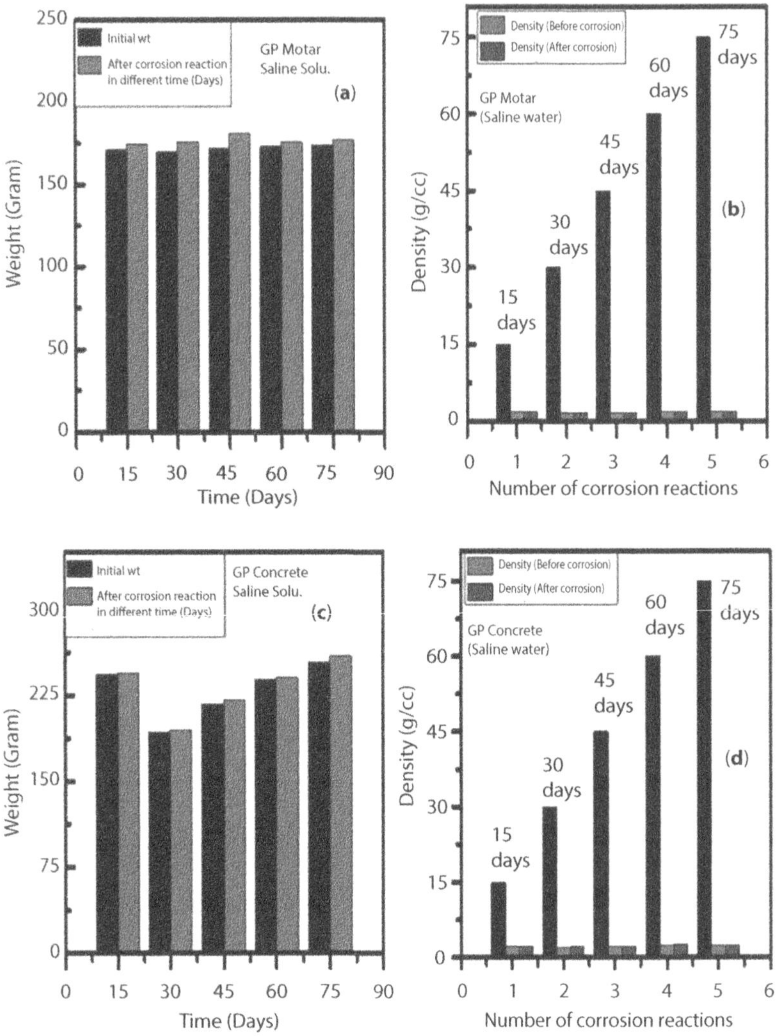

Figure 7.4 Comparison of weight change and bulk density between unexposed and exposed PA-based GP Mortar Concrete samples in saline water media.

change in weight and bulk density of exposed materials if compared with unexposed materials. Thus, the materials are found to be resistant to H_3PO_4 acid media.

Figures 7.7a and 7.7b shows bar charts of unexposed and exposed (for different lengths of exposure time) PA-based GP Mortar sample(s) in HCl acid media. Similar plots of bar charts for GP concrete (exposed in HCl acid media) are shown in Figures 7.7c and 7.7d. There is not much change in

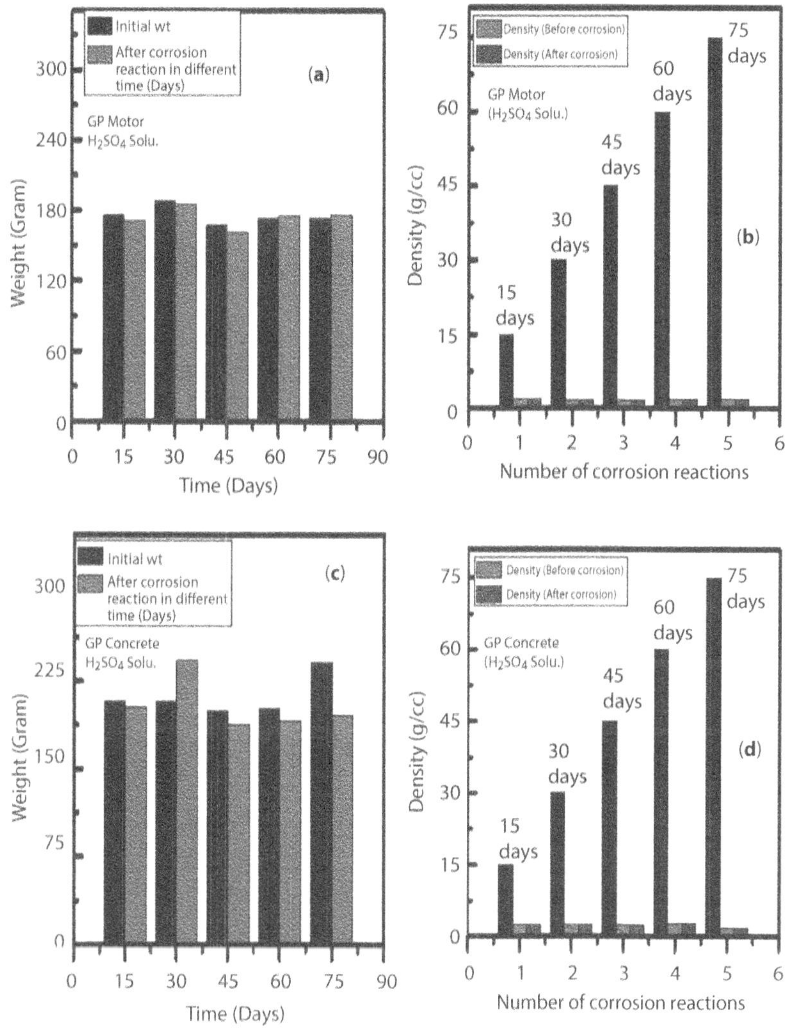

Figure 7.5 Comparison of weight change and bulk density between unexposed and exposed PA-based GP Mortar/Concrete samples in H_2SO_4 acid media.

weight and bulk density of exposed materials if compared with unexposed materials. Thus, the materials are found to be resistant to HCl acid media.

The mechanical property of unexposed and exposed samples (GP mortar/GP concrete) is shown in Table 7.3. Average strength of GP mortar and GP concrete do not change even after exposure in different media, i.e., normal water, distilled water, saline water, and HCl-acid media. This has helped to conclude the stability of the materials again.

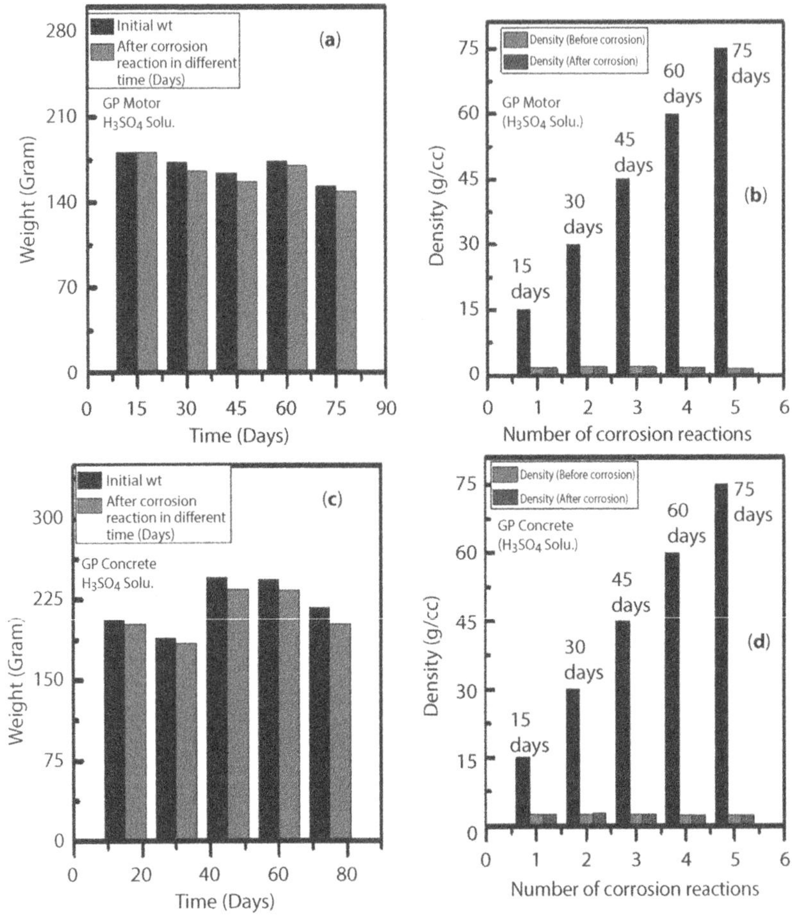

Figure 7.6 Comparison of weight change and bulk density between unexposed and exposed PA-based GP Mortar/Concrete samples in H_3PO_4 water media.

The mechanical property of GP mortar/GP concrete was evaluated at ambient temperature. In order to find the performance of two materials at higher temperatures, i.e., 70 °C, 80 °C, 90 °C, modulus of resilience (MOR) was determined. Tables 7.4 and 7.5 show the compressive strength value of GP mortar and GP concrete respectively. For accuracy, replicates were made (S_1, S_2, S_3), where subscripts 1, 2, 3 represented samples from each batch. In all cases, strength values decrease by 20% if compared with original strength value (24–19.5 MPa at higher test temperature). However, the mechanical strength of GP mortar did not change much at different temperatures.

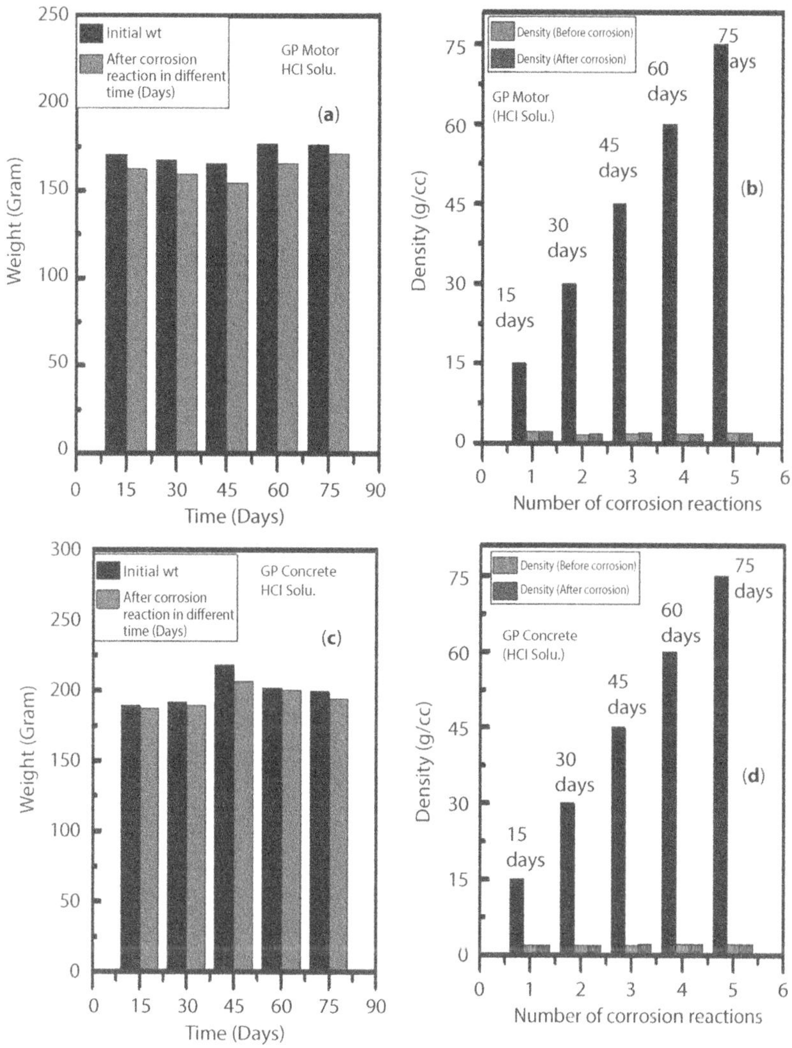

Figure 7.7 Comparison of weight change and bulk density between unexposed and exposed PA-based GP Mortar/Concrete samples in HCl acid media.

In order to confirm resolution of any material from test sample to the different media used for the corrosion study, liquid solution was collected from the container and chemicals were analyzed by atomic absorption spectroscopy (AAS). All the media were prepared with distilled water to minimize elemental content in the water used for preparation of solution. Owing to corrosion, the media normally contains soluble products from the

Table 7.3 Compressive strength (after exposure) of as-prepared PA-based GP mortar/concrete.

S. no.	Types of as-prepared GP materials	Av. compressive strength (before exposure, MPa)	Exposing media	Av. compressive strength (after exposure, MPa)
1	PA–GP Mortar [Component ratio of Mortar =1:3]	**24**	Normal water	24.1
			Distilled water	24.2
			HCl acid media	24.3
			Saline water media	24.1
2	PA–GP Concrete [Component ratio of Concrete=1:1.5:3]	**29.7**	Normal water	30
			Distilled water	29.9
			HCL acid media	30.1
			Saline water media	29.9

corroded samples. This will indirectly indicate the deterioration of materials after interacting with media. Because of this, the media was checked for change in the composition. Table 7.6 shows the chemical analyses of media. There is no marked presence of elements which are expected to emerge from the materials interacting with media. This helped us to again conclude the stability of GP concrete under different corrosive media.

The SEM images of the microstructure of exposed samples are shown in Figure 7.8. All the materials were exposed for 75 days in different media. For comparison purpose, the images shown in Figure 7.8 are marked with different letters, i.e., a, b, c, d, e, f, g and h. Topography features of all

Table 7.4 MOR values of PA-based GP mortar evaluated at temp. 70 °C, 80 °C, 90 °C.

S. no.	Components (85%)					
	Types of materials	PA (%)	FA (%)	TM (°C)	CS (MPa)	Av. CS (MPa)
1	S_1	21.25	63.75	70	19.0	19.50
2	S_2				19.2	
3	S_3				18.9	
4	S_1			80	20.3	19.76
5	S_2				19.7	
6	S_3				19.3	
7	S_1			90	19.2	19.30
8	S_2				19.2	
9	S_3				19.5	

**Note:* TM (testing temperature), CS (compressive strength), FA (fine aggregate).

Table 7.5 MOR values of PA-based GP concrete at temp. 70 °C, 80 °C, 90 °C.

	Components (85%)					
Types of materials	PA (%)	FA (%)	GA (%)	TM (°C)	CS (MPa)	Av. CS (MPa)
S_1	15.45	**23.18**	**46.36**	70	29.6	**29.7**
S_2					29.9	
S_3					29.8	
S_1				80	30.2	**30.2**
S_2					30.4	
S_3					30.0	
S_1				90	28.9	**29**
S_2					29.1	
S_3					29.0	

*Note: TM (testing temperature), CS (compressive strength), FA (fine aggregate), GA (granule aggregates).

Table 7.6 Atomic absorption study of three liquid samples collected from corrosive medium of PA-based GP mortar/concrete samples.

Test parameters												
Sample	**Fe**	**Cr**	**Ti**	**Ni**	**As**	**B**	**Ca**	**Mg**	**Se**	**Pb**	**Zn**	**Cd**
ID	ppm	ppm	ppm	ppm	ppm	ppm	(%)	(%)	ppm	ppm	ppm	Ppm
S1	18.0	<1	1.7	<1	0.3	10.0	0.0019	0.0014	<1	<1	<1	<1
S2	7.9	<1	<1	<1	1.0	22.0	0.011	0.009	<1	<1	<1	<1
S3	25.7	6.4	23.0	2.6	0.3	13.0	0.037	0.037	2.0	<1	6.9	<1

S1: Distilled water-75days (PA based GP Mortar), S2: saline water-75D (PA based GP Concrete), and S3: HC1 media-75D (PA based GP Concrete).

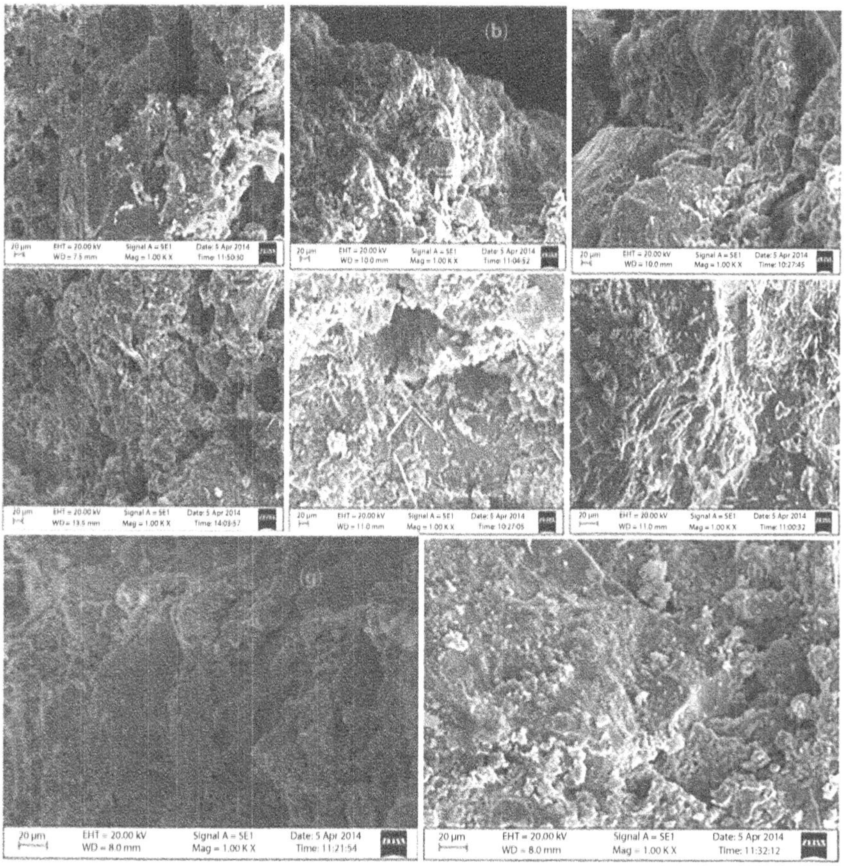

Figure 7.8 SEM images of PA-based GP mortar/concrete samples in different media: (a) HCl SOL-M-75D; (b) Distilled H_2O-M-75D; (c) Normal H_2O-M-75D; and (d) Saline H_2O-M-75D; (e) HCl SOL-C-75D; (f) Saline H_2O-C-75D; (g) Distilled H_2O-C-75D; (h) Normal H_2O-C-75D.

mortar samples are similar in nature. Rough facets are seen where powders are agglomerated due to geopolymerization of minerals. Widespread particles of different sizes are seen in the background. Particles have varying sizes ranging between 1–5 μm. There are some hollow regions in between reacted products. However, topological features of all samples are similar irrespective of media (Figures 7.8a–d). Some features can be marked in GP concrete which are not seen in PA-based GP mortar. The SEM image of GP concrete shows needle-shaped crystalline phase in the matrix; and river patterns are also seen in the matrix. It is evident from these images that there are some reactions occurring during the polymerization process.

Figures 7.9a and 7.9b shows the XRD pattern of exposed GP mortar/concrete samples in different corrosive media at room temperature. The objective of an XRD diffraction study is to investigate the occurrence of different phases of exposed samples (GP mortar/concrete). Intense peaks are found in the XRD pattern (Figures 7.9a & 7.9b), which reflects the presence of quartz and mullite. Similar phases are observed in the pond ash [11]. Mullite phase has a stable crystalline structure at ambient conditions. Mullite helps to retains strength and stability of the structure [65]. Semi-crystalline sodium aluminosilicate hydrate (N-A-S-H) phase is observed in the exposed mortar/concrete samples. It is a binding phase in geopolymer samples. In geopolymer samples, sodium plays an important role in balancing the charges of $[AlO_4]$ tetrahedrons. Occurrence of this phase increases the stability of geopolymer samples.

A crystalline phase of hydro sodalite ($Na_4Al_3Si_3O1_2OH$) is found in exposed samples. Zeolite underwent recrystallization as nepheline ($NaAlSiO_4$) and albite ($NaAlSiO_8$) hydrous aluminosilicate are also found after corrosive media treatments [66, 67]. The presence of all the phases helps to improve the strength and stability of the samples.

The FTIR spectra for pond ash-based geopolymer pastes [11] are shown in Figure 7.10. The main feature of the FTIR spectra is the central band occurring around 1093 cm^{-1}, which is attributed to Si–O–Si or Al–O–Si

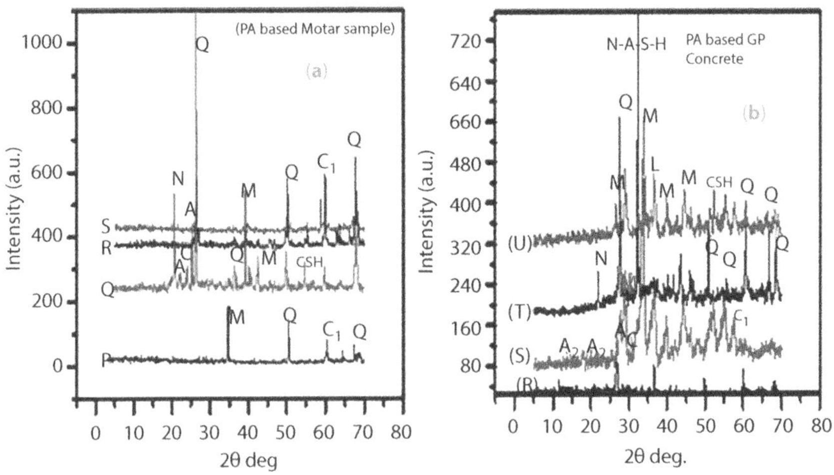

Figure 7.9 X-ray diffraction spectra of PA geopolymer-based mortar (a) and Concrete (b) samples after exposure to various corrosive environmental treatments [N: Nephelite, Q: Quartz, M: Mullite, A1: Anorthite, A2: Albite, C: Calcite, C1: Corundum, NASH: Sodium aluminium silicate hydrate, CSH: Calcium silicate hydrate, I: Illite].

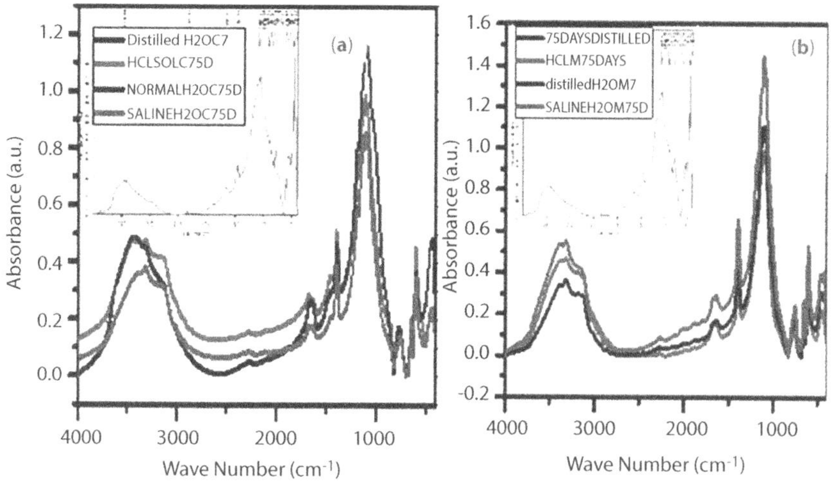

Figure 7.10 FTIR spectra of (a) exposed concrete samples and (b) mortar samples in different corrosive media (distilled water, normal water, saline water, and 1M HCl solution).

asymmetric stretching band. The main spectral band originally appearing in pond ash at about 1078 cm^{-1} has shifted to lower frequencies after geopolymerization took place. The larger the shift, the higher the degree of penetration of Al from the glassy part of the pond ash into the $(SiO4)^{4-}$ net. This indicates that the geopolymerization process was influenced by both parameters. Significant broad bands are observed at approximately 3450 cm^{-1} and 1640 cm^{-1} for the O–H stretching mode and O-H bending mode.

Figures 7.10a and 7.10b shows the FTIR spectra of exposed GP mortar/concrete in different corrosive environments (distilled water, normal water, saline water, and 1M HCl solution). After 75 days, all the samples were removed from different media and kept in a hot oven (1 h at 120 °C). Small amounts of samples were taken for FTIR studies. In these analyses, different peaks or bands were observed. The peak at 3610 cm^{-1} represents the stretching vibration of free O-H group, whereas peaks at 2980 and 2880 cm^{-1} signify stretching vibration of hydrogen-bonded O-H group. The peak at 1640 cm^{-1} represents the bending vibration of hydrogen-bonded O-H group [68–72]. The peak observed at 2360 cm^{-1} indicates C-O band, which represents stretching mode of vibration of C-O band. This is an indication of formation of CO_2 in the geopolymer mortar/concrete [68–72]. Peaks occurring at 1460 and 1390 cm^{-1} signify stretching vibration of C-O bond, which is related to the sodium carbonate Na_2CO_3 [68–72]. The broadest and strongest peak at 1020 cm^{-1} is the asymmetric stretching vibration of the Si-O-T bond (where T denotes Si or Al) [68–72]. The peak at 795 cm^{-1} is symmetric stretching vibration of the Si-O-Si bond [69]. Finally, the peaks at 776 and 694 cm^{-1} represent the crystalline phase of quartz components [68–73]. It should be noted that similar peaks or band, *viz.*, free O-H, hydrogen bonded O-H, C-O, Si-O-T and Si-O-Si bond, were found in unexposed geopolymer mortar/concrete.

7.3 Conclusions

A thorough characterization of geopolymer (GP) mortar/concrete samples was made after exposing them to corrosive media. This was aided by the use of advanced equipment such as an X-ray diffractometer (XRD) with a scanning electron microscope (SEM) attached, Fourier-transform infrared (FTIR) spectrometer, atomic absorption spectrometer (AAS), and ultimate strength machine (UTM). Cured GP materials showed evidence of geopolymerization. Subsequent experiments were conducted on GP mixed with sand and GP mixed with gravel to produce GP mortar and GP concrete. This was necessary because one of the aims of the study was to replace conventional cement with GP for construction purposes. As previously stated, these materials were also characterized by sophisticated equipment. Also, in the research work, corrosion behavior of pond ash-based geopolymer concrete as well as mortar samples were studied, which kept prepared samples in several media, like normal water, distilled water, and saline water; and in various acidic media like H_2SO_4 solution, HCl solution, and H_3PO_4 solution. The density changes and the weight changes were noted and the results were very interesting. Mechanical properties, such as compressive strength, morphology, and thermal stability, were well supported by the stability of GP concrete in different corrosive media.

Acknowledgments

The authors convey their sincere thanks to the Ministry of Mines, Government of India, for providing the grant to carry out this work [Grant number= F.No.:14/54/20214-Met.-IV dated: 29.12.2014]. The authors convey their sincere thanks to GIET, Gunupur, Rayagada, Odisha, India for providing lab facilities to do the research work. The authors would also like to thank the CRF, IIT Kharagpur, for providing testing facilities.

References

1. M. Schneider, M. Romer, M. Tschudin, and H. Bolio, (2011), Sustainable Cement Production-Present and Future, Cem. Concr. Res. Vol. 41 pp. 642-650 (http://dx.doi.org/10.1016/j.cemconres.2011.03.019).
2. D.A. Salas, A.D. Ramirez, C.R. Rodríguez, D.M. Petroche, A.J. Boero, and J. Duque-Rivera, (2016), Environmental Impacts, Life Cycle Assessment and Potential Improvement Measures for Cement Production: A Literature Review, J. Clean. Prod. Vol. 113, pp. 114-122 (http://dx.doi.org/10.1016/j.jclepro.2015.11.078).
3. M. Babaee, and A. Castel, (2016), Chloride-Induced Corrosion of Reinforcement in Low Calcium Fly Ash-Based Geopolymer Concrete, Cem. Concr. Res. 88, pp. 96-107, (http://dx.doi.org/10.1016/j.cemconres.2016.05.012).
4. B.W. Jo, J.S. Choi, K.W. Yoon, and J.H. Park, (2012), Material Characteristics of Zeolite Cement Mortar, Constr. Build. Mater. Vol. 36, pp. 1059-1065 (http://dx.doi.org/10.1016/j.conbuildmat.2012.06.061).
5. C. Gunasekara, D.W. Law, S. Setunge, and J.G. Sanjayan, (2015), Zeta Potential, Gel Formation and Compressive Strength of Low Calcium Fly Ash Geopolymers, Constr. Build. Mater. Vol. 95, pp. 592-599 (http://dx.doi.org/10.1016/j.conbuildmat.2015.07.175).
6. P. Duan, C. Yan, W. Zhou, and W. Luo, (2016), Fresh Properties, Mechanical Strength and Microstructure of Fly Ash Geopolymer Paste Reinforced with Sawdust, Constr. Build. Mater. Vol. 111, pp. 600-610 (http://dx.doi.org/10.1016/j.conbuildmat.2016.02.091).
7. A. Petrillo, R. Cioffi, C. Ferone, F. Colangelo, and C. Borrelli, (2016), Eco-Sustainable Geopolymer Concrete Blocks Production Process, Agric. Agric. Sci. Procedia, Vol. 8, 408-418 (http://dx.doi.org/10.1016/j.aaspro.2016.02.037).
8. K.H. Yang, J.K. Song, and K. Il Song, (2013), Assessment of CO_2 Reduction of Alkali-Activated Concrete, J. Clean. Prod. Vol. 39, pp. 265-272 (http://dx.doi.org/10.1016/j.jclepro.2012.08.001).
9. J. Davidovits, (1991), Geopolymers, J. Therm. Anal. Vol. 37, pp. 1633-1656 (http://dx.doi.org/10.1007/BF01912193).
10. C. Monticelli, M.E. Natali, A. Balbo, C. Chiavari, F. Zanotto, S. Manzi, and M.C. Bignozzi, A Study on the Corrosion of Reinforcing Bars in Alkali-Activated Fly Ash Mortars Under Wet and Dry Exposures to Chloride Solutions, Cem. Concr. Res. Vol., 87, (2016) pp. 53-63 (http://dx.doi.org/10.1016/j.cemconres.2016.05.010).
11. M. Panigrahi, S. Mohanty, R.R. Dash, and R.I. Ganguly, (2018), Development of Novel Constructional Material from Industrial Solid Waste as Geopolymer, IOP Conf. Series: Materials Science and Engineering, Vol. 410, pp. 01-12 (doi:10.1088/1757-899X/410/1/012012).

12. P. Sukmak, S. Horpibulsuk, S.L. Shen, P. Chindaprasirt, and C. Suksiripattanapong, (2013), Factors Influencing Strength Development in Clay-Fly Ash Geopolymer, Constr. Build. Mater. Vol. 47, pp. 1125-1136 (http://dx.doi.org/10.1016/j.conbuildmat.2013.05.104).
13. W. Mozgawa, J. Deja, (2009), Spectroscopic Studies of Alkaline Activated Slag Geopolymers, J. Mol. Struct. Vol. 924-926, pp. 434-441 (http://dx.doi.org/10.1016/j.molstruc.2008.12.026).
14. R.H. Haddad, and O. Alshbuol, (2016), Production of Geopolymer Concrete using Natural Pozzolan: a Parametric Study, Constr. Build. Mater. Vol. 114, pp. 699-707 (http://dx.doi.org/10.1016/j.conbuildmat.2016.04.011).
15. M. Najimi, J. Sobhani, B. Ahmadi, and M. Shekarchi, (2012), An Experimental Study on Durability Properties of Concrete Containing Zeolite as a Highly Reactive Natural Pozzolan, Constr. Build. Mater. Vol. 35, pp. 1023-1033, (http://dx.doi.org/10.1016/j.conbuildmat.2012.04.038).
16. B.H. Mo, H. Zhu, X.M. Cui, Y. He, and S.Y. Gong, (2014), Effect of Curing Temperature on Geopolymerization of Metakaolin-based Geopolymers, Appl. Clay Sci. Vol. 99, pp. 144-148 (http://dx.doi.org/10.1016/j.clay.2014.06.024).
17. M. Sarkar, K. Dana, and S. Das, (2015), Microstructural and Phase Evolution in Metakaolin Geopolymers with Different Activators and Added Aluminosilicate Fillers, J. Mol. Struct. Vol. 1098 pp. 110-118 (http://dx.doi.org/10.1016/j.molstruc.2015.05.046).
18. M.H. Cornejo, J. Elsen, C. Paredes, and H. Baykara, (2015), Hydration and Strength Evolution of Air-Cured Zeolite-Rich Tuffs and Siltstone Blended Cement Pastes at Low Water-To-Binder Ratio, Clay Miner. Vol. 50, pp. 133-152 (http://dx.doi.org/10.1180/claymin.2015.050.1.12).
19. M. Kr_ol, J. Minkiewicz, W. Mozgawa, M. Krol, J. Minkiewicz, and W. Mozgawa, (2016), IR Spectroscopy Studies of Zeolites in Geopolymeric Materials Derived from Kaolinite, J. Mol. Struct. Vol. 1126, pp. 200-206 (http://dx.doi.org/10.1016/j.molstruc.2016.02.027).
20. H. Baykara, M.H. Cornejo, R. Murillo, A. Gavilanes, C. Paredes, J. Elsen, (2017), Preparation, Characterization and Reaction Kinetics of Green Cement: Ecuadorian Natural Mordenite-Based Geopolymers, Mater. Struct. Vol. 50, pp. xxx-xxx, (http://dx.doi.org/10.1617/s11527-017-1057-z).
21. J. Davidovits, (2002), Environmentally Driven Geopolymer Cement Applications in Geopolymer 2002 Conf. Vol. xxx, pp.1-9.
22. J. Davidovits, (2017), Mineral Polymers and Methods of Making them, US4349386 A, 1982 (https://www.google.com/patents/US4349386, 12 February 2017).
23. P. Duxson, A. Fernandez-Jimenez, J.L. Provis, G.C. Lukey, A. Palomo, and J.S.J. van Deventer, (2007), Geopolymer Technology: the Current State of the Art, J. Mater. Sci. Vol. 42, pp. 2917-2933 (http://dx.doi.org/10.1007/s10853-006-0637-z).

24. E. Rakanta, T. Zafeiropoulou, and G. Batis, (2013), Corrosion Protection of Steel with DMEA-based Organic Inhibitor, Constr. Build. Mater. Vol. 44, pp. 507-513 (http://dx.doi.org/10.1016/j.conbuildmat.2013.03.030).
25. A. Ali, and S. Frantisek, (2005), Sulfuric Acid Attack on Hardened Paste of Geopolymer Cements, Part 1. Mechanism of Corrosion at Relatively High Concentrations", Ceramics–Silikáty, Vol. 49, pp. 225-229.
26. Ali, A., and Frantisek, S., Sulfuric Acid Attack on Hardened Paste of Geopolymer Cements, Part 2. Corrosion Mechanism at Mild and Relatively Low Concentrations, Ceramics-Silikaty, (2006), Vol. 50, pp.1-4.
27. X.J. Song, M. Marosszeky, M. Brungs, and R. Munn, (2005), Durability of Fly Ash based Geopolymer Concrete Against Sulphuric Acid Attack", 10 DBMC International Conferences on Durability of Building Materials and Components, Lyon, France, 17- 20 April 2005
28. S.E. Wallah and B.V. Rangan (2006), Low Calcium Fly Ash based Geopolymer Concrete: Long Term Properties, Research Report GC 2, Curtin University of Technology, Australia.
29. R.N. Thakur and S. Ghosh, (2007), Fly ash based Geopolymer composites, Proceedings of 10th NCB International seminar on Cement and building materials, New Delhi, India Nov, Vol.3, pp.442- 451.
30. A. Palomo, M.W. Grutzeck, and M.T. Blanco, (1999), Alkali Activated Fly Ashes a Cement for the Future, Cement and Concrete Research, Vol. 29, pp.1323-1329.
31. A. Ali, and S. Frantisek, (2001), Nitric Acid Attack on Hardened Paste of Geopolymeric Cements Part 1, Ceramics-Silikáty, Vol. 45, pp. 81-88.
32. A. Ali, and S. Frantisek, (2001), Nitric Acid Attack on Hardened Paste of Geopolymeric Cements, Part 2, Ceramics-Silikaty, Vol. 45, pp. 143-149.
33. C. Shia, and J.A. Stegemann, (2000), Acid Corrosion Resistance of Different Cementing Materials", Cement and Concrete Research, Vol. 30, pp. 803-808.
34. E. Rakanta, T. Zafeiropoulou, and G. Batis, (2013) Corrosion protection of steel with DMEA-based organic inhibitor, Constr. Build. Mater. Vol. 44, pp. 507-513. (http://dx.doi.org/10.1016/j.conbuildmat.2013.03.030).
35. P. Faustino, A. Bras, and T. Ripper, (2014), Corrosion Inhibitors' Effect on Design Service Life of RC structures, Constr. Build. Mater. Vol. 53, pp. 360-369 (http://dx.doi.org/10.1016/j.conbuildmat.2013.11.098).
36. A.A. Sagües, (2004), Galvanized Steel Reinforcement in Concrete, in: Galvaniz. Steel Reinf. Concr., pp. 71-86 (http://dx.doi.org/10.1016/B978-008044511-3/50018-9).
37. Z. Ahmad, (2006), Concrete Corrosion, Princ. Corros. Eng. Corros. Control, Vol. 12, pp. 609-646 (http://dx.doi.org/10.1016/B978-075065924-6/50013-1).
38. M. Büchler, (2005), Corrosion Inhibitors for Reinforced Concrete, Elsevier Ltd, 2005, (http://dx.doi.org/10.1533/9781845690434.190).
39. J.M. Gaidis, (2004), Chemistry of Corrosion Inhibitors, Cem. Concr. Compos. Vol. 26, pp. 181-189 (http://dx.doi.org/10.1016/S0958-9465(03)00037-4).

40. M. Forsyth, and M.Z. Lourenco, (1997), Corrosion and protection of steel in concrete, Corros. Mater. Vol. 22, pp. 13-16 (http://dx.doi.org/10.1533/9781845693398.136).
41. M. Ormellese, M. Berra, F. Bolzoni, and T. Pastore, (2006), Corrosion Inhibitors for Chlorides Induced Corrosion in Reinforced Concrete Structures, Cem. Concr. Res. Vol. 36, pp. 536-547 (http://dx.doi.org/10.1016/j.cemconres.2005.11.007).
42. T.A. S€oylev and M.G. Richardson, (2008), Corrosion inhibitors for steel in concrete: stateof-the-art report, Constr. Build. Mater. Vol. 22, pp. 609-622 (http://dx.doi.org/10.1016/j.conbuildmat.2006.10.013).
43. V. Saraswathy and H.W. Song, (2007), Improving the Durability of Concrete by using Inhibitors, Build. Environ. Vol. 42, pp. 464-472 (http://dx.doi.org/10.1016/j.buildenv.2005.08.003).
44. E.S. Dewi and J.J. Ekaputri, (2017), The Influence of Plain Bar on Bond Strength of Geopolymer Concrete, AIP Conference Proceedings, Vol. 1855, pp. 030017-xxx.
45. E.A. Azimi, M.M. Al Bakri Abdullah, L.Y. Ming, H.C. Yong, K. Hussin, and I.H. Aziz, (2016), Processing and Properties of Geopolymers as Thermal Insulating Materials: A Review-Rev. Adv. Mater. Sci., Vol. 44, pp. 273-285.
46. Z.F. Farhana, H. Kamarudin, A. Rahmat and A.M. Mustafa Al Bakri, (2013), A Study on Corrosion Behavior of Reinforcement Bar Embedded in Geopolymer Paste by Open Circuit Potential, Australian Journal of Basic and Applied Sciences, Vol. 7, pp. 230-235.
47. S. Mundraa, S.A. Bernala, M. Criadoa, P. Hlaváčekb, G. Ebellb, S. Reinemannb, G.J.G. Gluthb, and J.L. Provisa, (2017), Steel Corrosion in Reinforced Alkali-Activated Materials, RILEM Technical Letters, Vol. 2, pp. 33-39.
48. A. Allahverdi and F. Škvára, (2005), Sulfuric Acid Attack on Hardened Paste of Geopolymer Cements Part 1. Mechanism of Corrosion at Relatively High Concentrations, Ceramics-Silikáty, Vol. 49, pp. 225-229.
49. R. Manickavasagam and G. Mohankumar, (2017), Durability Studies on the High Calcium Flyash based GPC, International Journal of Engineering, Science and Technology, Vol. 9, pp. 1-9.
50. Y.P. Asmara (2016), Long Term Corrosion Experiment of Steel Rebar in Fly Ash-Based Geopolymer Concrete in NaCl Solution, International Journal of Corrosion Volume 2016, pp. 1- 5.
51. T. Udhaya Kumar, M, Jayadurgalakshmi, (2019), Non-Destructive Behavior of Corroded Reinforced Geopolymer Concrete Beams, International Journal of Engineering and Advanced Technology (IJEAT), Vol. 9, pp. 5411-5414.
52. T.B. Aulia, M. Muttaqin, M. Afifuddin, and M. Zaki, (2020), Effect of Utilizing Geopolymer Fly Ash on Potential and Corrosion Rate of Reinforcement in High-Strength Concrete, IOP Conf. Series: Materials Science and Engineering, Vol. 933, pp. 012047-xxx.

53. G. Rambabu, D.A. Naidu, and D. Venkateswarlu, (2019), Experimental Developments on Corrosion Resistivity of Low Calcium Fly Ash-Based Geopolymer Concrete, International Journal of Recent Technology and Engineering (IJRTE), Vol. 8, pp. 5320-5327.
54. R. Bayuaji, S. Darmawan, N.A. Husin, B. Wibowo, S. Subekti, M.M. Al Bakri Abdullah, and T.R. Biyanto, (2017), The Effect of Corrosive Environment on Geopolymer Concrete Tensile Strength, MATEC Web of Conferences 97, 01036
55. P. Bhardwaj, R. Gupta, D. Mishra, S.K. Sanghi, S. Verma, and S.S. Amritphale, (2020), Corrosion and Fire Protective Behavior of Advanced Phosphatic Geopolymeric Coating on Mild Steel Substrate, *Silicon*, 12, 487–500.
56. W. Yodsudjai, (2020), Corrosion Behavior of Reinforcement in Concrete with Different Compositions, Journal of Sustainable Cement-Based Materials: Vol xxx, pp. 1-5.
57. S. Sikora, E. Gapys, B. Michalowski, T. Horbanowicz, and M. Hynowski, (2018) Geopolymer Coating as a Protection of Concrete Against Chemical Attack and Corrosion, SOLINA 2018 E3S Web of Conferences, Vol. 49, pp. 00101-08.
58. Faiz U.A. Shaikh, (2014), Effects of Alkali Solutions on Corrosion Durability of Geopolymer Concrete, Advances in Concrete Construction Vol. 2, pp. 109-123.
59. A.R. Lakshmi Narayanan, and K. Vasugi, (2019), Study of Corrosion of Steel Bars in GGBS based Geopolymer Concrete, International Journal of Innovative Technology and Exploring Engineering (IJITEE), Vol. 8, pp. 1747-1752.
60. T. Udhaya Kumar, and M. Jayadurgalakshmi, (2018), Studies on Corrosion of Reinforced Geopolymer Concrete Beams, International Journal of Civil Engineering and Technology (IJCIET), Vol. 9, pp. 1348-1356.
61. B. Mahboubi, Z. Guo, and Hao Wu, (2019), Evaluation of Durability Behavior of Geopolymer Concrete Containing Nano-Silica and Nano-Clay Additives in Acidic Media, J. Civil Eng. Mater. App., Vol. 3, pp. 163-171.
62. N.U. Auqui, H. Baykara, A. Rigail, M.H. Cornejo, and J.L. Villalba, (2017), An investigation of the effect of migratory type corrosion inhibitor on mechanical properties of zeolite-based novel geopolymers, Journal of Molecular Structure, Vol. 1146, pp. 814-820.
63. H.A. Ariza-Figueroa, J. Bosch, M.A. Baltazar-Zamora, R. Croche, G. Santiago-Hurtado, L. Landa-Ruiz, J.M. Mendoza-Rangel, J.M. Bastidas, F. Almeraya-Calderón, and D.M. Bastidas, (2020), Corrosion Behavior of AISI 304 Stainless Steel Reinforcements in SCBA-SF Ternary Ecological Concrete Exposed to $MgSO_4$, Materials, Vol. 13, pp. 2412-24128.
64. B.S. Mohammed, S. Haruna, M.M.A. Waha, M.S. Liew, and A. Haruna, (2019) Mechanical and Microstructural Properties of High Calcium Fly Ash One-Part Geopolymer Cement Made with Granular Activator, Heliyon, Vol. 5, pp. 1-9.

65. H. Schneider, J. Schreuer, and B. Hildmann, (2008), Structure and Properties of Mullite - A Review, Journal of the European Ceramic Society, Vol. 28, pp. 329-344.
66. Y. Jun, S. Yoon, and J. Eun Oh, (2017), A Comparison Study for Chloride-Binding Capacity between Alkali-Activated Fly Ash and Slag in the Use of Seawater, *Appl. Sci.,* Vol. 7, pp. 971-14 (https://doi.org/10.3390/app7100971).
67. D. Moro, R. Fabbri, J. Romano, G. Ulian, A. Calafato, A. Solouki, C. Sangiorgi, and G. Valdrè, (2021), Thermal, X-ray Diffraction and Oedometric Analyses of Silt-Waste/NaOH-Activated Metakaolin Geopolymer Composite, J. Compos. Sci., Vol. 5, pp. 269-13 (https://doi.org/10.3390/jcs5100269).
68. F. Mádai, F. Kristály, G. Mucsi, (2015), Microstructure, Mineralogy and Physical Properties of Ground Fly Ash based Geopolymers, Ceramics – Silikáty, Vol. 59, pp. 70-79
69. C.A. Rees, J.L. Provis, G.C. Lukey, and J.S.J. van Deventer, (2007), Attenuated Total Reflectance Fourier Transform Infrared Analysis of Fly Ash Geopolymer Gel Aging, *Langmuir,* Vol. 23, pp. 8170-8179.
70. T. Bakharev, (2005), Resistance of geopolymer materials to acid attack, Cement and Concrete Research, Vol. 35, pp. 658-670.
71. H.A. Abdel-Gawwad, S.A. Abo-El-Enein, (2014), A novel method to produce dry geopolymer cement powder, Housing and Building National Research Center HBRC Journal, Vol. xxx, pp. xxx-xxx. (http://dx.doi.org/10.1016/j.hbrcj.2014.06.008)
72. H.E.E. Fouad, W.H. Soufi, A.S. Elmannaey, M. Abd-El-Aziz, and H. El-Ghazaly, (2020), Durability and Steel Corrosion Resistance of Slag with Metakaolin based Geopolymer Concrete, Journal of Engineering and Applied Science, Vol. 67, pp. 1381-1398.
73. T.A. Aiken. J. Kwasny, and W. Sha, (2020), Resistance of Fly Ash Geopolymer Binders to Organic Acids, Materials and Structures, 53:115, pp. 1-18, (https://doi.org/10.1617/s11527-020-01549-x).

8

Applications, Challenges and Opportunities of Geopolymer Materials

Ashis Kumar Samal[1], Muktikanta Panigrahi[2*], Ratan Indu Ganguly[3] and Radha Raman Dash[4]

[1]Department of Civil Engineering, Gandhi Institute of Engineering and Technology University, Gunupur, India
[2]PG Department of Materials Science, Maharaja Sriram Chandra Bhanja Deo University, Keonjhar Campus, Odisha, India
[3]Department of Metallurgical Engineering, National Institute of Technology, Raurkela, Odisha, India
[4]CSIR-National Metallurgical Laboratory, Jamshedpur, Jharkhand, India

Abstract

With their unique and favorable performance, geopolymer materials have attracted extensive attention in recent years. They combine some of the characteristics of inorganic polymers, cements and ceramics due to their special polycondensed network structure. This chapter introduces the different applications of geopolymer that have recently been developed, with the promise of even more applications to come in the future. Some of their applications in civil engineering include innovative binders, toxic waste encapsulation, heat-resistant coatings and adhesives, novel high-temperature ceramic materials, and structure repair. In addition to the opportunities that geopolymer materials present, the existing challenges faced by experienced construction engineers are also put forward, many of which can be met by this kind of materials. Therefore, it is desirable to extend the use of these materials for future requirements.

Keywords: Geopolymer, pavements, fire-resistant materials, ceramics, insulating materials, protective coatings, challenges, and opportunities

**Corresponding author*: muktikanta2@gmail.com

Muktikanta Panigrahi, Ratan Indu Ganguly and Radha Raman Dash (eds.) Development of Geopolymer from Pond Ash-Thermal Power Plant Waste: Novel Constructional Materials for Civil Engineers, (227–240) © 2023 Scrivener Publishing LLC

8.1 Introduction

The term "geopolymer" was first introduced by Davidovits in 1978 [1]. Geopolymer materials are characterized by molecular chains and networks similar to organic polymers [2], which are formed by the action of alkali molecules on inorganic compounds bearing silicon and aluminium atoms. The strength of these materials (i.e., GP) is due to chain formation as well as network formation (crosslinking like thermosetting polymer). These materials are found suitable to replace conventional ordinary Portland cement (OPC) [3]. Since these materials can be produced from industrial waste such as fly ash, pond ash, and steel plant slag, they are very cost-effective and will also help reduce pollution. As such, manufacturing of cements is an energy-intensive process, releasing dust into the atmosphere which is injurious to human health and the environment.

Geopolymer can also be called alkali-activated cement or inorganic polymer cement. However, the development of these GP materials is still in its infancy. These inorganic products consist of $[SiO_4]^{-4}$ and $[AlO_4]^{-4}$ tetrahedral networks. The influences of alumino-silicate materials, such as metakaolin and low-calcium fly ash, were investigated by many researchers [2–7], and their properties are considered suitable for reformation of aggregate similar to ordinary Portland cement (OPC). Based on past results, high-calcium fly ash has been successfully used as a raw material in the geopolymer mixture [8, 9]. A wide range of applications for geopolymer materials [10–15] in the fields of new ceramics, binders, and matrices for hazardous waste stabilization, require fire-resistant materials and high-tech materials, which are emerging trends in current research.

Geopolymer technology is primarily used to reduce the environmental impact of ordinary Portland cement. However, geopolymer materials have various other areas of application, from the field of civil engineering to automobile and aerospace industries, as briefly discussed below.

Geopolymer concrete, bricks, paver blocks, etc., are an innovative civil engineering material in the construction sector [16]. Geopolymer concrete is produced by chemical treatment of silica-alumina bearing molecules. Fly ash is a silica-alumina-bearing waste material which is obtained from coal combustion in thermal power plants worldwide. Geopolymer concrete is produced using geopolymer paste, sand and gravels. Related civil engineering materials are shown in Figure 8.1.

In the present scenario, there is growing demand for sustainable binders or binder systems such as geopolymers and related alkali-activated aluminosilicate. Geopolymer binders are already being marketed as special

Figure 8.1 Civil engineering geopolymer materials: (a) GP-based bricks, (b) GP-based fired bricks, (c) GP-based paver block, (d) GP-based concrete, and (e) GP-based brick wall.

products, and are explored for use in structural products such as mortar and concrete. Geopolymer binders are environmentally friendly and are called green materials. They have the potential to replace cement in construction sectors. Geopolymer binders are mainly prepared utilizing aluminosilicate-bearing materials. Waste materials, such as fly ash (FA), pond ash (PA), granulated blast furnace slag (GBFS), rice husk ash and red mud, etc., mostly contain alumina and silica in major quantities. Alumina and silica are converted to aluminosilicate during geopolymer binder production. Such binder can reduce the carbon dioxide emissions about 80–90% if used [16, 17]. Geopolymer as well cement is suitable for storage wells under deep down-hole stress conditions and carbon capture owing to its good acid resistance, high mechanical strength, durability and low permeability [18]. Because of their potential behavior, GP cements are used in the place of OPC cement [19, 20].

Another potential application of geopolymer is protective coatings (PCs). Geopolymer protective coatings (GPCs) are prepared from industrial wastes that contain a high-volume of alumina and silica precursors with treatment of alkali-activator solutions. GPCs are cluster materials. They can be potentially used to protect the surfaces of civil infrastructures. Geopolymer coating is a network structured material. The network is formed through the aluminium tetrahedral and silicon tetrahedra unit.

It is a kind of inorganic coating which is low cost, fire proof, water proof, and nontoxic to the environment, etc. Geopolymer with different proportions of β-phase of SiC are used to prepare the GP-type paints. It has better corrosion resistance and wear resistance than virgin geopolymer coatings [21–23]. Also, the reflective heat insulation performance of GP coating is improved by the addition of talcum powder, hollow glass microspheres and sericite powder during the synthesis of geopolymer from metakaolin and sodium silicate solutions [23]. Hybrid coating materials are prepared using GP concrete and alkaline-activated metakaolin/fly ash for chloride-induced corrosion protection [23]. Geopolymer-based coating products are shown in Figure 8.2.

Geopolymer is an inorganic polymer which is ceramic in nature. It is shaped below 100 °C. Geopolymer chains or networks contain mineral molecules and are connected with covalent bonds [24–26]. Raw materials of geopolymer ceramics are mainly minerals of geological origin. Geopolymer technology offers a new method for direct preparation of the ceramics. Geopolymer ceramics can form a structure through high sintering temperatures.. The process is convenient for preparing different complex-shaped GP ceramic products. They have good overall performance which is comparable to inorganic ceramics. Some of the geopolymer-based ceramic products are shown in Figure 8.3.

Different sources, such as modern industry, agriculture, waste disposal, etc., produce waste incineration residue containing harmful heavy metals. Geopolymer technology can convert industrial solid waste bearing aluminosilicate minerals into value-added products due to their flexibility and capability to immobilize and stabilize the wastes inside the GP chains or networks. Metals like Co, Cu, Pb, Cd, Ni, Zn, Pd, As, Ra and U can be incorporated into the GP network, thereby decreasing the heavy

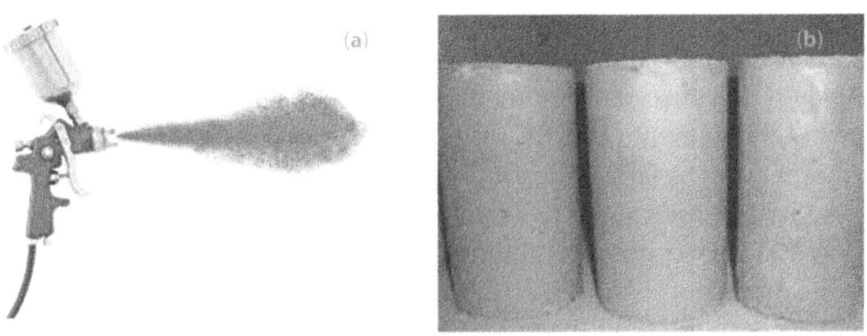

Figure 8.2 Geopolymer-based coatings: (a) Metakaolin-based GP coatings and (b) GP coating on concrete.

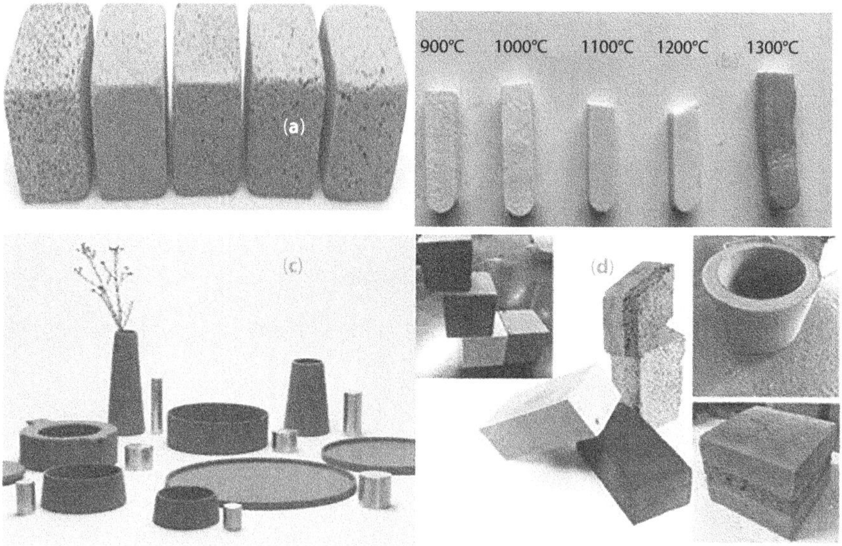

Figure 8.3 Geopolymer ceramic products: (a) ceramic waste-based GP foam, (b) fired ceramic GP materials, and (c & d) ceramic GP decorative products.

ions mobility through precipitation, substitution or physical encapsulation [27, 28].

In the present scenario, air pollution is mainly due to industrial production, natural disasters, engineering construction, automobile exhaust, etc. Consequently, building structures will become dirtier. This issue has forced the scientific community and industries to develop viable alternative materials. One of those viable products in development are GP-based self-cleaning concrete materials. Due to their unique properties, the performance of self-cleaning GP concrete is improved by the introduction of photocatalytic materials, i.e., ZnO and TiO_2 [29–31].

Ongoing research currently involves geological, chemical and archaeological aspects of studying the stability and chemical make-up of archaeological structure. Currently, there is historical documentation of geopolymeric cements used to make archaeological structures with improved durability. Examples of archaeological structures made with geopolymeric materials are shown in Figure 8.4.

Aircraft technology requires low-cost, environmentally friendly, fire-resistant matrix composite materials [32–36]. For this technology, essential flammability of materials is 50 kW/m, which makes them resistant to incident heat flux of a fully developed aviation fuel fire penetrating a cabin opening before the fire spreads to the cabin compartment. The objective is

Figure 8.4 Use of geopolymeric materials in archaeological structures.

to remove the threat of cabin fire in order to avoid aircraft accidents. Newly developed aircraft materials with low to moderate cost are mostly applicable in transportation and infrastructure areas where a high degree of fire resistance is needed. New low-cost geopolymer-based materials are being used in aircraft. Carbon/geopolymer composite-based materials are now tested and used as aircraft materials due to their outstanding properties. A geopolymer-composite exhaust pipe system has been developed. Some fire-resistant materials are shown in Figure 8.5.

Generally, infrastructure like bridges is made up of steel-reinforced concrete, which is corroded by corrosives such as salt water and deicers. Concrete and brick structures are repaired by externally bonding with flexible fiber sheet composites. Also, organic matrix-based continuous fiber composites are used in infrastructure to wrap concrete columns.

Figure 8.5 Geopolymer fire-resistant materials.

Due to their flammability behavior, the use of organic polymer matrix and fiber-reinforced composites limits their use in offshore oil platforms, military vehicles and public transportation where fire endurance and fire hazards are important considerations. Their susceptibility to fire currently limits their use. Carbon fiber-reinforced geopolymer composites do not ignite, burn, or release any smoke even after extended heat flux. Carbon fiber-reinforced geopolymer matrix-based composite retains 63% of its original 245 MPa flexural strength even after a simulated large fire.

Engineers face many difficulties when it comes to effectively insulating buildings with minimum energy loss. When designing insulating materials, their safety, resistance to fire, and environmental friendliness must be kept in mind. Alkali-activated mineral materials, i.e., geopolymers, can be produced as foamed materials [37–40] that have low weight, low thermal conductivity, high fire resistance, and exceptional mechanical strength properties [37–40]. Due to the above excellent properties, geopolymer foams have found many applications in areas of industry as well construction [37–40]. The trade name of commercially available geopolymer foam materials is TROLIT, which was the first geopolymer foam [41, 42]. Thermal conductivity value of this material is 0.037 W/mK. It is fabricated using blowing agents. Hydrogen peroxide (H_2O_2) or sodium perborate ($NaBO_3$) is used as blowing agent [43–45]. The density of geopolymer foams ranges between 0.2 and 0.8 g/cm^3. An apparent heat resistance and thermal conductivity is found to be 1200 °C and 0.037 W/mK, respectively. To develop foamed geopolymer with the best possibility of insulation performance, reduction of coefficient thermal conductivity is necessary to decrease the density of foam material. This is possible by increasing the amount of foaming agent [43–45]. Bell and Kriven [46] developed foams with the addition of H_2O_2 and aluminum powder and investigated their performance. Vaou and Panias [47] prepared geopolymer foam and investigated the coefficient of thermal conductivity (0.030 W/mK). The strength property plays a key role in many potential applications. Two- or three-layer foamed geopolymer partitions are gradually required for insulation and construction purposes.

Other Applications of Geopolymers

Because of their excellent properties and potential applicability, geopolymers have also been applied to make environmentally friendly adhesive (for wood-based panels and modern biotechnology), high-tech carbon fiber materials (used inside and outside of aircraft), dielectric materials (for electronic packaging) and molecular sieve catalysts material, etc.

8.2 Challenges

World is experiencing environmental pollution due to generation of industrial wastes, by-products, gases, etc. Waste produced from thermal power plants and other sources cause global warming. This provides an opportunity for scientists, engineers, and researchers to accept the challenge of solving the pollution problem. One way to do this is by inventing new-generation materials, which will not need thermal processing. In addition, waste generation can be minimized by controlling industrial pollution. In other words, a carbon-free world needs to be created in order to enhance the quality of human life.

Cement is a very important construction material in the 21st century century. Production of cement is a heat-intensive process which increases global warming as well as pollution. If cement can be replaced by other materials for construction purposes, then pollution can be controlled to a great extent. This will reduce the unwanted gases that cause pollution.

The work currently be done has attempted to make new materials, i.e., geopolymer, for construction purposes. As stated earlier, many scientists are working on geopolymer production through fly ash and other resource materials.

8.3 Opportunity

Replacement of cement by geopolymer will be a boon to the global economy and health of the world's population. It will also encourage entrepreneurial activities, thereby helping to solve the unemployment problem. In addition, geopolymer composites will immensely benefit the modern world due to their ease of production, which will reduce the availability and cost of raw materials. This will help the entire world develop at a faster pace. In conclusion, some of the benefits of geopolymer materials are enumerated below:

i. Since raw materials are available as waste, geopolymer is a cost-effective material.
ii. Utilization of waste materials will protect land.
iii. Simplicity of its production process, where lower temperature is needed.
iv. Ease of production.
v. Replacement of costlier conventional cement.

vi. Solves employment problem by increasing entrepreneurial activities, which will help to develop nations.

8.4 Conclusions

Among the advantages of geopolymer are its easy availability, ease of manufacturing, resource- and energy-saving properties, and excellent properties similar to organic polymers, ceramics and cement. Therefore, many researchers are paying more and more attention to them. In this chapter, geopolymer and its applications were broadly reviewed. Based on their performance, the raw materials of geopolymers will continue to enhance applications, and some new potential applications will be found. Based on the reviews, opportunities in future research work can be recommended.

Acknowledgments

The authors convey their sincere thanks to the Ministry of Mines, Government of India, for providing the grant to carry out this work [Grant number= F.No.:14/54/20214-Met.-IV dated: 29.12.2014]. The authors also convey their sincere thanks to GIET, Gunupur, Rayagada, Odisha, India, for providing lab facilities to do the research work. The authors would also like to thank the CRF, IIT Kharagpur, for providing testing facilities.

References

1. J. Davidovits, (2008), Geopolymer Chemistry and Applications (Geopolymer Institute, Saint-Quentin, France).
2. H. Xu and J.S.J. Van Deventer, (1999), The Geopolymerization of Natural Aluminosilicates, Proceedings: 2nd International Conference on Geopolymere"99, (Geopolymer Institute), pp. 43-63.
3. A. Palomo, M.W. Grutzeck, and M.T. Blanco-Varela, (1999), Alkali Activated Fly Ashes: Cement for the Future, Cement and Concrete Research, Vol.xxx, pp. 1323-1329.
4. H. Xu and J.S.J. Van Deventer, (2002), Geopolymerization of Multiple Minerals, Minerals Engineering, Vol. xxx, pp. 1131-1139.
5. A.M. Fernandez-Jimenez, E.E.A. Lachowski. Palomo, and D.E. Macphee, (2004), Microstructural Characterization of Alkali-activated PFA Matrices for Waste Immobilization, Cement and Concrete Composites, Vol. xxx, pp. 1001-1006.

6. T. Bakharev, (2005), Geopolymeric Materials Prepared using Class F Fly Ash and Elevated Temperature Curing, Cement and Concrete Research, Vol. xxx, pp. 1224-1232.
7. S. Andini, R. Cioffi, F. Colangelo, T. Grieco, F. Montagnaro, and L. Santoro, (2008), Coal Fly Ash as Raw Material for the Manufacture of Geopolymer-based Products, Waste Management, Vol. xxx, pp. 416-423.
8. P. Chindaprasirt, T. Chareerat, and V. Sirivivatnanon, (2007), Workability and Strength of Coarse High Calcium Fly Ash Geopolymer, Cement and Concrete Composites, Vol. xxx, pp. 224-229.
9. A. Sathonsaowaphak, P. Chindaprasirt, and K. Pimraksa, (2009), Workability and Strength of Lignite Bottom Ash Geopolymer Mortar, Journal of Hazardous Materials, Vol. xxx, pp. 44-50.
10. J.G.S. Van-Jaarsveld, J.S.J. Van Deventer, and L. Lorenzen, (xxx), The Potential Use of Geopolymeric Materials to Immobilize Toxic Metals" Part I. Theory and Applications. Mineral Engineering, Vol. 10, pp. 659-669.
11. W.M. Kriven, J.L. Bell, and M. Gordon, (2004), Geopolymer Refractories for the Glass Manufacturing Industry, Ceramic Engineering Science Process, Vol. 25, pp. 57-79.
12. I. Lancellotti, E. Kamseu, M. Michelazzi, L. Barbieri, A. Corradi, and C. Leonelli, (2010), Chemical Stability of Geopolymers Containing Municipal Solid Waste Incinerator Fly Ash, Waste Management, Vol. 30, pp. 673-679.
13. S. Andini, R. Cioffi, F. Colangelo, C. Ferone, F. Montagnaro, and L. Santoro, (2010), Characterization of Geopolymer Materials Containing MSWI Fly Ash and Coal Fly Ash,. Advanced Science and Technology, Vol. 69, pp. 123-128.
14. J.W. Giancaspro, C.G. Papakonstantinou, and P. Balaguru, (2010), Flexural Response of Inorganic Hybrid Composites with E-Glass and Carbon Fibers, Journal of Engineering Materials and Technology, Vol. 132, pp. 1-8.
15. C. Ferone, F. Colangelo, R. Cioffi, F. Montagnaro, and L. Santoro, (2011), Mechanical Performances of Weathered Coal Fly Ash based Geopolymer Bricks, Procedia Engineering, Vol. 21, pp. 745-752.
16. R. McCaffery, (2002), Climate Change and the Cement Industry, Global Cement and Lime Magazine, Environment-Special Issue, Vol.xxx, pp. 15-19.
17. J. Davidovits, (1999), Chemistry of Geopolymeric Systems, Terminology (Proceeding of Geopolymer"99 International Conference, Saint-Quentin, France).
18. B.V. Rangan, (2008), Studies on Fly Ash-Based Geopolymer Concrete, Malaysian Construction Research Journal, Vol. 3, pp. xxx-xxx.
19. B.B. Jindal, D. Singhal, S. Sharma, and Parveen, (2018), Enhancing Mechanical and Durability Properties of Geopolymer Concrete with Mineral Admixture, Computers and Concrete, Vol. 21, pp.1-10 (DOI: https://doi.org/10.12989/cac.2018.21.3.000).

20. Ganesan Lavanya and Josephraj Jegan, (2015), Durability Study on High Calcium Fly Ash Based Geopolymer Concrete, Advances in Materials Science and Engineering, Vol. 2015, pp.1-8 (http://dx.doi.org/10.1155/2015/731056).
21. J. Temuujin, A. Minjigmaa, W. Rickarda, M. Leea, I. Williamsa, and A.Van Riessena, (2010), Fly ash based Geopolymer Thin Coatings on Metal Substrates and its Thermal, Evaluation Journal of Hazardous Materials, Vol.180, pp. 748-752.
22. J. Temuujin, A. Minjigmaa, W. Rickard, M. Lee, I. Williams, and A.V. Riessen, (2009), Preparation of Metakaolin based Geopolymer Coatings on Metal Substrates as Thermal Barriers, Applied Clay Science, Vol.46, pp. 265-270.
23. Y. Mao, L. Biasetto and P. Colombo, (2020), Metakaolin-based Geopolymer Coatings on Metals by Airbrush Spray Deposition, Journal of Coatings Technology and Research, Vol. 17, pp. 991-1002.
24. L. Alves, A. Nogueira1, J. dos Santos and S. de Barros, (2019), A Quick Overview on Geopolymer Chemistry and General Properties, Res Dev Material Sci. Vol. 12, pp. 1249-1251 (DOI: 10.31031/RDMS.2019.12.000781).
25. J. Davidovits, Geopolymer Chemistry and Applications, 5th ed., pp. 1-680 (ISBN: 9782954453118).
26. J. Davidovits, (1988) Geopolymer Chemistry and Properties. Proceedings of the 1st International Conference on Geopolymer '88, Vol. 1, Compiegne, 1-3 June 1988, pp. 25-48.
27. K. Kupwade-Patil, E.N. Allouche, Md. R. Islam, and A. Gunasegaram, (2014), Encapsulation of Solid Waste Incinerator Ash in Geopolymer Concretes and Its Applications, ACI Materials Journal, V. 111, pp. 691-702. (DOI: 10.14359/51686834).
28. P. Cong and Y. Cheng, (xxx): Advances in Geopolymer Materials: A Comprehensive Review, J. Traffic Transp. Eng. (Engl. Ed.) Vol. xxx, pp. 1-32 (https://doi.org/10.1016/j.jtte.2021.03.004).
29. S.N. Zailan, N. Mahmed, M.M. Al Bakri Abdullah, and A.V. Sandu, (2016), Self-cleaning geopolymer concrete-A review, IOP Conf. Series: Materials Science and Engineering, Vol. 133, pp.1-8 (doi:10.1088/1757-899X/133/1/012026).
30. S.N. Zailan, A. Bouaissi, N. Mahmed, and M.M. Al Bakri Abdullah, (2019), Influence of ZnO Nanoparticles on Mechanical Properties and Photocatalytic Activity of Selfcleaning ZnOBased Geopolymer Paste, Journal of Inorganic and Organometallic Polymers and Materials, Vol. xxx, pp. 1-11. (https://doi.org/10.1007/s10904-019-01399-3).
31. K.U.A. Sanalkumar and E.-H. Yang, (2021), Self-Cleaning Performance of Nano-TiO2 Modified Metakaolin-based Geopolymers, Cement and Concrete Composites, Vol. 115, pp.1-12 (https://doi.org/10.1016/j.cemconcomp.2020.103847).
32. M. Welter, M. Schmücker, K.J.D. MacKenzie, (2015), Evolution of the Fibre-Matrix Interactions in Basalt-Fibre-Reinforced Geopolymer-Matrix Composites after Heating, J. Ceram. Sci. Tech., Vol. 06, pp.17-24 (DOI: 10.4416/JCST2014-00034).

33. P.R. Jackson, T.A. Parthasarathy, A. Ross, and D.W. Radford, (2018), Use of Interphase in Geopolymer Matrix Composites for Improved Toughness, Ceramics International, (DOI: https://doi.org/10.1016/j.ceramint.2018.10.076).
34. G. Roviello, L. Ricciotti, C. Ferone, F. Colangelo, and O. Tarallo, (2015), Fire Resistant Melamine based Organic-Geopolymer Hybrid Composites, Cement & Concrete Composites, (DOI: http://dx.doi.org/10.1016/j.cemconcomp.2015.03.007).
35. H.C. Yong, L.Y. Ming, M.M. Al Bakri Abdullah, and K. Hussin, (2015), Fire Resistant Properties of Geopolymers: A Review, Key Engineering Materials Vol. 660, pp 39-43 (DOI:10.4028/www.scientific.net/KEM.660.39).
36. R. Lyon, (1996), Fire Response of Geopolymer Structural Composites, pp. 1-10 (DOT/FAA/AR-TN95/22).
37. G. Roviello, C. Menna, O. Tarallo, L. Ricciotti, F. Messina, C. Ferone, D. Asprone, and R. Cioffi, (2017), Lightweight Geopolymer-based Hybrid Materials, Composites Part B, (DOI: 10.1016/j.compositesb.2017.07.020).
38. L. Ricciotti, A. Occhicone, A. Petrillo, C. Ferone, R. Cioffi, and G. Roviello, (2019), Geopolymer-based Hybrid Foams: Lightweight Materials From A Sustainable Production Process, Journal of Cleaner Production (https://doi.org/10.1016/j.jclepro.2019.119588).
39. R.M. Novais, R.C. Pullar, and J.A. Labrincha, (2019), Geopolymer Foams: An Overview of Recent Advancements, Progress in Materials Science, (DOI: https://doi.org/10.1016/j.pmatsci.2019.100621).
40. M. Łach, K. Korniejenko, and J. Mikuła, (2016), Thermal Insulation and Thermally Resistant Materials Made of Geopolymer Foams, Procedia Engineering, Vol. 151, pp. 410-416 (DOI:10.1016/j.proeng.2016.07.350).
41. M. Łach, J. Mikuła, W.-T. Lin, P. Bazan, B. Figiela, and K. Korniejenko, (2020), Development and Characterization of Thermal Insulation Geopolymer Foams Based on Fly Ash, Proceedings of Engineering and Technology Innovation, Vol. 16, pp. 23-29
42. M. Łach, (2021), Geopolymer Foams-Will They Ever Become a Viable Alternative to Popular Insulation Materials? A Critical Opinion, Materials, Vol. 14, pp. 1-15 (https://doi.org/10.3390/ma14133568)
43. V. Ducman, and L. Korat, (2016). Characterization of geopolymer fly-ash based foams obtained with the addition of Al powder or H_2O_2 as foaming agents. Materials Characterization, Vol. xxx, pp. xxx-xxx (DOI:10.1016/j.matchar.2016.01.019).
44. Z. Abdollahnejad, Z. Zhang, H. Wang, and M. Mastali, (2017), Comparative Study on the Drying Shrinkage and Mechanical Properties of Geopolymer Foam Concrete Incorporating Different Dosages of Fiber, Sand and Foam Agents, High Tech Concrete: Where Technology and Engineering Meet, Vol. xxx, pp. 42-48 (https://link.springer.com/chapter/10.1007/978-3-319-59471-2_6)

45. Z. Abdollahnejad, F. Pacheco-Torgal, T. Félix, W. Tahri, and J. Barroso Aguiar, (2015), Mix Design, Properties and Cost Analysis of Fly Ash-Based Geopolymer Foam, Construction and Building Materials, Vol. 80, 18-30.
46. J.L. Bell and W.M. Kriven, (2009), Preparation of Ceramic Foams from Metakaolin-Based Geopolymer Gels, Ceramic Engineering and Science Proceedings, Vol. 29, pp. 96-111 (DOI:10.1002/9780470456200.ch10).
47. V. Vaou and D. Panias, (2010), Thermal Insulating Foamy Geopolymers from Perlite, Minerals Engineering, Vol. 23, pp. 1146-1151 (DOI:10.1016/j.mineng.2010.07.015).

Index

Accelerated durability testing, 55–56
Accelerated mortar bar test, 53
Agate mortar, 206
Air-cooled BF slag (ACBFS), 42
Aircraft, geopolymer-based materials in, 231, 232
Air pollution, 29–30, 32, 170, 231
Akrotiri, 3
Albite, 218
Alccofine powder, 93
Alkali-activated-based geopolymer, 153
Alkali-activated FA/BFS, 153
Alkali-activated materials (AAMs), 199
Alkali activation of ceramic waste (AACW), 50, 76
Alkali dosage effects, 155
Alkaline sodium silicate, 48
Alkali silicon-oxoaluminate, 45
Alumina, 151, 153
Aluminium salt solution, 171–172
Alumino-silicate geopolymer, 45, 47, 49, 75, 201
Aluminosilicate hydrogels, 49, 75
Aluminum and silicon-bearing materials, GP preparation, 154
Aluminum powder, 174, 233
Aluminum silicates, activation of, 196
Amazonian kaolin, 171
Anji Bridge (China), 8
Anthracite, 26
Archaeological structures, with geopolymeric materials, 231, 232
Argon gas, 102, 103
Argon oxygen decarburization (AOD) slag, 48, 73
Asbestos in roofing components, 171
Ash/ashes, types, 18–33
  cigar ash, 19
  coal ash and fly ash, 21–24
  coconut shell ash, 21
  fly ash generation, 24, 25
  pond ash, 29–31
  pond ash management, importance of, 32–33
  pulverized fuel ash, various uses of, 31–32
  quarry dust, 20
  RHA, 19
  thermal power plant ashes, nature and composition, 24, 26–29
  volcanic ash, 19–20
  wood ash, 19
Atomic absorption spectroscopy (AAS), 213, 216

Backfilling material (BM), microstructure of, 154
Bagasse cellulose fiber, 173
Baghdad Power Plant, 52, 77
BALCO, 170
Bamboo fibers, 171
Bands (peaks), 219–220
Basalt fibers, 174–175

Basalt strands, 172
Basic oxygen furnace (BOF) slag, 40, 42
Beauvais Cathedral, 11
BET (Brunauer, Emmett and Teller) surface areas of pond ashes, 93
BFS. *see* Blast furnace slag (BFS)
Binders, geopolymer, 228–229
Biochar, 48, 74
Bituminous coal, 26
Blaine air-permeability method, 35
Blast furnace slag (BFS), 40, 41, 42, 44
  alkali-activated, binding mechanism of, 153
  compressive strength tests, 153
  flexural strength test, 153
  geopolymer production from, 73–77
  GGBFS, 42, 49, 50, 74–76
Blowing agent, 233
BOF (basic oxygen steelmaking) slag, 42
Bottom ash, 23, 24, 29, 30, 92, 124–125
  chemical composition of, 125
  collection, 92
  engineering properties, 125
Bragg's law, 101
Brick(s), 12, 13, 125, 229, 232
  brick-making technology, 9
  burnt clay, 29
  eco-friendly, 48, 74, 174
  fired, 6, 8, 23
  manufacture of, 31
  mold-made mud, 2
  mud-brick, 2, 4, 5
  production, 16
  weight loss of, 174
  with lime mortar, 9
Bridges, organic matrix-based continuous fiber composites, 232, 233
Bronze Age, 3
Brown coal fly ash, 94
Bulk density, PA-based GP mortar/ concrete, 207–213
Burnt clay bricks, production of, 29

Calcined kaolin, 48
Calcium aluminate cements, 49, 74
Calcium aluminum silicate hydrate (CASH), 153
Calcium electric arc ferronickel slags, 73, 76
Calcium hydroxide, 202
Calcium silicate hydrate (CSH), 153, 156
Carbon dioxide (CO_2), emission of, 73, 152, 196
Carbon fiber, 174–175, 233
Carbonic acid, 198
Carbon negative cement, 52, 78
Carl Zeiss Supra-40 Scanning Electron Microscope, 206
Casting,
  PA-based geopolymer products, 203, 204, 205
  PA/HCFC slag-based GP preparation, 157
  PA/jute fiber-based GP, 179–181
  preparation of geopolymer from PA, 99, 100
Cast iron, use of, 16
Caustic soda. *see* Sodium hydroxide
Cellulose, 171
Cement(s); *see also* High carbon ferrochrome (HCFC) slag-based GP cementitious materials; Pond ash (PA)-based GP cementitious materials; Pond ash (PA)–jute fiber-based GP cementitious materials
  -based materials, corrosion of, 196, 198, 199
  alkali-activated, 47, 72, 228
  biochar with, 48, 74
  blended, 124

calcium aluminate, 49, 74
carbon negative, 52, 78
cementing agent, 26, 27
FABC, 49, 75
factories, 9
flowable slurry along with, 32
for storage, 229
formulation of, 49
hydrates and aggregate, 197
in acid media, 201
in concrete, 73
industry, slag in, 40, 41
inorganic polymer, 72
magnesium chloride phase of, 77
magnesium oxy chloride (sorel)-based, 52, 77
magnesium oxy sulfate-based, 52, 77
mortar, durability behavior of, 55
PC. *see* Portland cements (PC)
pore radii in, 197
production, 46, 72, 92, 196, 234
replacement of, 29, 47, 48, 74, 124, 125, 148, 154, 161, 200, 202, 207, 228, 229, 234
resource material for preparing, 31
RHA with, 19
SCMs and, 200
seawater resistant, 55
self-cementing behavior, 27
slag, 42, 44
sodium sulfate-activated slag, 51, 77
SSC, 49, 74
thermal energy, 53
using ferrochrome slag (FS), 155
Ceramics, geopolymer, 230, 231
Characterization of geopolymer (GP), 155, 159, 171, 172, 197
DSC. *see* Differential scanning calorimetry (DSC)
FTIR. *see* Fourier-transform infrared spectroscopy (FTIR)
IR, 100
of prepared samples, 127
PA-based GP mortar/concrete (before and after) corrosion, 206–207
PA/HCFC-based geopolymeric material, 158–159
pond ash, 32
SEM. *see* Scanning electron microscopy (SEM)
techniques, 48, 49, 74
TGA. *see* Thermogravimetric analysis (TGA)
XRD. *see* X-ray diffraction (XRD)
Chartres Cathedral, 11
China,
ancient, 8–9
coal deposits in, 21, 23
Chloride-induced corrosion durability of GP, 201
Chopped fibers, 179, 181, 186
Cigar ash, 19
Cinder. *see* Slag(s)
Class C fly ash, 26–29
Class F fly ash, 26–28, 155
Clinker lumps, 29
Coade stone, 16
Coal,
combustion, 23–25
consumption, 21, 23
fly ash slurry spill, 30
fossil fuel, 21
Coal ash(es), 21–24, 29–31
chemical composition, 36
chemical properties of, 35–36
compaction behavior, 36–37
leaching behavior, 37–38
mineralogical phases, 36
morphology of, 33
permeability, 37
physical characteristics of, 33–38
solubility of solids, 36
specific gravity of, 33
specific surfaces of, 35

strength behavior, 37
surface area of, 33
utilization, 38–39
Coal fly ash, for preparation of GP, 154
Coconut shell ash, 21
Coir fibers, residual, 171
Compacted mass, formation, 141
Compaction, 33–34
behavior, of coal ash, 36–37
PA/HCFC slag-based GP preparation, 157
PA-based geopolymer products, 203–205
PA/jute fiber-based GP, 179–181
preparation, PA-based GPs, 99, 100
Compression testing,
of as-prepared geopolymer samples, 127
PA-jute fiber-based GP, 172
Compressive strength,
of corroded geopolymer composite, 200–201
of PA/HCFC slag GP materials, 153, 154, 156, 158–160
PA-based GP mortar/concrete, 206, 212, 214, 215
PA-jute fiber-based GP, 173, 174, 182–186
Concrete,
cracking in, 54
deterioration process of, 53–54
dissolution mechanism of steel in, 54
durability of, 53–55
electrochemical process, 54
mechanism of steel corrosion, 54
Construction materials, historical development of,
accelerated durability testing, 55–56
ash/ashes, types, 18–33
cigar ash, 19
coal ash and fly ash, 21–24
coconut shell ash, 21
fly ash generation, 24, 25
pond ash, 29–31
pond ash management, importance of, 32–33
pulverized fuel ash, various uses of, 31–32
quarry dust, 20
RHA, 19
thermal power plant ashes, nature and composition, 24, 26–29
volcanic ash, 19–20
wood ash, 19
chronological development, 2–18
ancient China, 8–9
ancient Egypt, 4–5
ancient Greece and Rome, 5–8
ancient Mesopotamia, 4
Copper Age and Bronze Age, 3
eighteenth century, 15–16
Iron Age, 3–4
Middle Ages, 9–11
Neolithic Age, 2–3
nineteenth century, 16–17
Renaissance, 11–15
seventeenth century, 15
Steel Age, 3–4
twentieth century, 17–18
coal ash(es),
physical characteristics of, 33–38
utilization, 38–39
durability of concrete, 53–55
deterioration process, 53–54
dissolution mechanism of steel in, 54
electrochemical process, 54
mechanism of steel corrosion, 54
geopolymers, 45–53
activated, preparation of, 48
alkali activation mechanism for, 45
at ambient temperature, 49–50
applications of, 47
by alkaline activation, 49

chemical composition of, 45
constituents of, 46–52
defined, 45
fabricated plant fiber-based, 48
fire-resistant, 51
GPC, 47
hybrid inorganic-organic, 50
in alkaline solution, 5
kaolin, 51
metakaolin-based, 51
non-colored and colored, 50
properties, 45, 52–53
reaction, schematic representation of, 46
sol-gel method for, 48
through alkali-silicate activation, 50
waste glass-based, 49
overview, 1–2
slag, 39–45
ACBFS, 42
BFS, 41, 42
BOF, 42
chemistry, 39
converter, 44
electric arc furnace slag, 44
ferrous and non-ferrous smelting processes, 39
GBFS, 42
generation of, 40–44
GGBFS, 42
iron and steel slag, 43, 44
LD or steel slag, 42
optical photo images, 41
properties and utilization, 44–45
reducing, 44
rock, 42
synthetic, 39
Converter slag, 44
Copper Age, 3
Copper mine tailings, 48, 74
Corchorus capsularis, 178
Corchorus olitorius, 178
Corrosion,
behavior in different corrosive medium, 53
chloride-induced corrosion durability of GP, 201
electrochemical process of, 199
HCl corrosive media, 53
inhibitors, 198
in M30 grade geopolymer concrete, 201
in sulphate and acid media, 200
mechanism of, 199
mechanism of steel, 54, 55
of OPC concrete, 198
of pond ash (PA)-based geopolymer products,
causes of, 197
characterizations, 206–207
chemicals and materials, 203
comparison of weight change and bulk density, 207–213
compositions of PA-based concrete mixtures, 205
compositions of PA-based mortar mixtures, 204
concrete preparation process, 204–205
curing process, 204, 205
experimental details, 203–220
mechanical property of GP mortar/GP concrete, 211, 212, 214, 215
molding, casting and compaction, 203–205
mortar preparation, 203–204
overview, 196–202
preparation of, 203–205
raw material preparation, 203, 205
results and discussion, 207–220
of steel reinforcement, 198
potential measurement, 200
property of steel-reinforced concrete structures, 55

rate, 200
reinforced bar coated with GP paste, 198
resistance of high-strength reinforced concrete, 200
types of, 198
Cotton, 171
Cracking in concrete structures, 54
Crack widths, 174
Crushed residual waste glass (RWG), 74
Curing,
compacted mass formation, 141
green materials, 202
microwave-oven, 93
of GP paste, 198
PA, preparation of geopolymer from, 100
PA-based geopolymer products, 204, 205, 207
PA/HCFC slag-based GP preparation, 157
PA/jute fiber-based GP, 174, 179–182
temperatures, 105, 106, 128–132, 207
thermal, 94
times, 128–132, 156, 207
length of, 106–109
variations of, 106, 107

Dan River coal ash spill, 30
Desiccator method, 35
Dicalcium silicate, 44
Differential scanning calorimetry (DSC),
characteristic properties of FP pastes, 141, 146, 147
HCFC slag-based GPs, 156, 158, 159
PA-based GP mortar/concrete, 207
PA-based GPs, 102–104, 112, 115
PA-jute fiber-based GP, 175, 182
Dog bone tensile testing, 174
Dolomite, 52
DSC. *see* Differential scanning calorimetry (DSC)
Dumped fly ash, 24
Durability,
accelerated durability testing, 55–56
chloride-induced corrosion durability of GP, 201
of AAMs, 199
of cement mortar, 55
of concrete, 53–55
test in sulfuric acid, 93

Eco-friendly GP composite, 48, 74, 171–172, 174
EDX. *see* Energy-dispersive X-ray (EDX) analysis
Egypt, ancient, 4–5
Eighteenth century, constructional materials, 15–16
Electric arc furnace (EAF) slag, 40, 44
Electron microscopy techniques, 101–102
Electron tomography, pores in geopolymer, 94
Embankments, design, 30
Energy-dispersive X-ray (EDX) analysis,
as-received PA and prepared PA-based GP, 141, 146
characterization of prepared samples, 108–110, 127
HCFC slag-based GPs, 158–161
PA-based GP mortar/concrete, 206
PA/jute fiber-based GP, 181
Environmental pollution, 46, 72, 125, 152, 234
Epoxy-based organic resins, 49
Ettringite, 51, 75, 77
Eucalyptus waste, 171

Fabricated plant fiber-based geopolymer, 48

Fall cone method, 35
Ferrochrome slag (FS). *see* High carbon ferrochrome (HCFC) slag-based GP
Field emission scanning electron microscope (FESEM),
  as-received pond ash sample, 141, 146
  characterization of prepared samples, 101–102, 127
  geopolymer cement using HCFA, 202
Fired clays, 6
Fire hazards, 233
Fire-resistant geopolymers, 51, 77
Fire-resistant matrix composite materials, 231, 232
Flame atomic absorption spectrometry (F-AAS), 206
Flammability, of GP materials, 231–233
Flexural strength,
  of PA/HCFC slag GP materials, 153–155
  PA-jute fiber-based GP, 173, 174
Florence cathedral, dome of, 14, 15
Flushing gas stream, fluctuations in, 102
Fly ash (FA), 21–24
  alkali-activated, binding mechanism of, 153
  brown coal, 94
  chemical composition of, 26–29
  classes of, 26–29
  compressive strength tests, 153
  different mass ratios of, 172
  dry density of, 37
  flexural strength test, 153
  for plant growth, 34
  generation, 24–26, 92
  low-calcium, 47, 72
  PFA, 31–32
  pH value, 36
  pond ash *vs.*, 30–31
  pozzolanic behavior of, 31
  solubility of solids, 36
  strength of, 36
Fly ash-based geopolymer, 76
Fly ash-based geopolymer concretes (FAGPC), 154
Fly ash belite cement (FABC), 49, 75
Foaming agent, 233
Fossil fuels, 21, 22
Fourier-transform infrared spectroscopy (FTIR),
  geopolymer cement using HCFA, 202
  HCFC slag-based GP, 156, 159
  nature of the geopolymeric reaction using, 154
  PA-based GP mortar/concrete, 206, 218–220
  PA-based GPs, 100, 113, 127
  PA-jute fiber-based GP, 175, 182, 186–188
Fragmentation, process of ash formation, 20
Free swell index, 34–35
FTIR. *see* Fourier-transform infrared spectroscopy (FTIR)

Gas adsorption method, pores in geopolymer, 94
Geopolymer concrete (GPC), 47, 78
Geopolymer foam concretes (GFCs), 50, 76
Geopolymer protective coatings (GPCs), 229–230
Geopolymers (GPs), 45–53
  applications of, 47
  chemical composition of, 45
  constituents of, 46–52
  defined, 45
  formation mechanism, 78–81
  materials,
    applications, 228–233

benefits of, 234–235
challenges, 234
opportunity, 234–235
overview, 228–233
strength, 228
mechanical properties of, 128, 141–147, 171, 202
micrographs of, 108, 109
overview, 72–78
parameters of, 78
preparation, from pond ash, 98–100, 126
properties, 45, 52–53
synthesis, 73–78
Glass fiber, 174
Glass for architectural purposes, 7–8
Glass panes, 16
Global warming, 46, 47, 72, 73, 196, 234
GPs. *see* Geopolymers (GPs)
Grain size distribution, 34
Granulated blast furnace slag (GBS or GBFS), 42, 44
Great Pyramid of Giza, 4
Great Wall of China, 8–9
Greece, ancient, 5–8
Greenhouse gas effect, 46, 47, 72, 73, 196
Ground granulated blast furnace slag (GGBFS), 42, 49, 50, 155
corrosion rate for, 201
geopolymer production from, 74–76, 154
Groundwater pollution, 32
Gypsum, 37, 49

HCFC. *see* High carbon ferrochrome (HCFC) slag-based GP cementitious materials
Heating, weight loss of GP cured samples and, 146, 147
Heat insulation performance of GP coating, 230
Hibiscus cannabinus, 178
High-calcium fly ash (HCFA), 202
High carbon ferrochrome (HCFC) slag-based GP cementitious materials,
experimental details, 156–163
characterizations of PA/HCFC-based geopolymeric material, 158, 159
PA/HCFC-based geopolymeric mortar and concrete, 158, 159
PA/HCFC slag-based GP preparation, 157
results and discussion, 159–163
source of materials, 156–157
overview, 152–156
Hollow glass microspheres, 230
Hydraulic conductivity, in soil, 34
Hydrogen peroxide (H_2O_2)
as blowing agent, 233
solution, 76
Hydro sodalite, crystalline phase of, 218

Ilmenite smelting, 39
Imhotep, 5
Index properties, in geotechnical engineering practice, 35
India,
coal ashes, 36
coal deposits in, 21, 23
demand for energy in, 32
steel industry in, 40
thermal power plants in, 124
Industrial wastes,
bearing aluminosilicate minerals, 230
dumping and landfilling, 170
GPC from, 229
GP production from, 72, 74, 76, 77, 228
Inhibitors, corrosion, 198
Insulating materials, 233
Ionic tetrahedral coordination, 79

Iron, use, 15, 16
Iron Age, 3–4, 6
Iron aluminum oxides, amorphous, 36
Iron Bridge at Coalbrookdale, 16
Iron slag, 40, 43, 44

Jade Gate Pass (Yumenguan), 9
Jayadurgalakshmi, M., 199–201
Jiayuguan's Great Wall, 9
Jones, Inigo, 15
Jute fiber(s), 175–176
  -based GP, PA and. *see* Pond ash (PA)–jute fiber-based GP
  chemical composition, 178
  physical properties of, 176–178
  uses of, 176, 177

Kaolin, 51, 76
Kaolinite, 76
King's College Chapel, 11
Kingston Fossil Plant, 30
Kizhi (Russia), church in, 12, 13
Korogho church in Georgia, 10, 11

Labor, in the Renaissance, 13
LD (Linz-Donawitz) slag, 42, 44, 45
Leaching behavior, 37–38
Lighthouse of Alexandria, 6
Lignite, 27
Lime reactivity, 36
Limestone fillers (F-Lime), 49, 74
Linz-Donawitz (LD) slag, 42, 44, 45
Loba Chemicals, 126, 175
Louvre in Paris, roof of, 16
Lye. *see* Sodium hydroxide

Magma, 20
Magnesium oxy chloride (sorel)-based cements, 52, 77
Magnesium oxy sulfate-based cements, 52, 77
Magnesium sulfate, 154
Manganese oxides, 36
Masonry techniques of ancient Greece and Rome, 5–6
Media, corrosive, 213, 214, 216t, 218
Menkaure Pyramid, 4
Merck (India), 126, 156–157, 175
Mercury porosimetry method, pores in geopolymer, 94
Merwinite, 49
Mesh sizes, variations of, 105
Mesopotamia, ancient, 4
Metakaolin, 47–49, 51, 72, 73, 76, 173
  -based geopolymeric mortar, 77
  synthesis of geopolymer from, 230
Metallurgical slag, 51, 77
M30 grade geopolymer concrete, corrosion in, 201
Microwave-oven curing, 93
Middle Ages, 9–11
Mineralogical phases, 36
Ming Dynasty Great Wall, 9
Minitab software, 127
Modulus of resilience (MOR), PA-based GP mortar, 212, 215
Molding,
  PA-based geopolymer products, 203–205
  PA/HCFC slag-based GP preparation, 157
  PA/jute fiber-based GP, 179–181
  preparation of geopolymer from PA, 99, 100
Monasticism, building techniques, 9
Mongolian pond ash, 93
Morphologies of geopolymer, 36, 94, 148, 158, 171, 182, 206
  as-received pond ash, 141, 146
  coal ash, 33
  pond ash, 31, 32
  surface, of prepared materials, 101
Mortar(s),
  PA/jute fiber-based GP, 178–185
  preparation, PA-based geopolymer, 203–204

Mud brick, 2, 4, 5
Mullite, 218
Museum of Prehistoric Thera in Santorini (Greece), 3

NALCO Navratna Company, 91, 98, 126, 156, 170, 175, 203
Nanchan Temple (Wutai), 8
Nardite, 49
Natural ennore sand, 93
Natural fibers, 170–173, 175–177
Natural pozzolan (PN), 49
Neolithic Age, 2–3
Neolithic tools, 2
Nepheline, 218
New Stone Age, 2–3
Nineteenth century, constructional materials, 16–17
Nishibayashi, S., 55–56
Nomograms, for mechanical properties of GP, 141–147
Non-destructive testing (NDT) technique, 100
Notre Dame Cathedral, Paris, 11
NTPC, 170

Oil, energy source, 21, 22
Oligomers, defined, 79
Olivine, 44
OPC. *see* Ordinary Portland Cement (OPC)
Open-circuit potential (OCP) method, for corrosion measurement, 200
Ordinary Portland Cement (OPC), 31, 46, 47, 53, 72, 228
- -based concrete, 154
- corrosion behavior of, 198
- environmental impact of, 228
- partial substitution of, 172
- production, 152, 196
- using HCFA, 202
- with SCBA, 202

Ores, defined, 39

PA. *see* Pond ash (PA)
Palatine Chapel, Aachen, 9
Pantheon in Rome, 6
Parthenon, 6
Passivity of embedded steel, 55
Paver blocks, geopolymer, 228, 229
Peaks (bands), 219–220
Peat wood, 171
Percussion cup method, 35
Perkin Elmer Pyris Diamond analyzer, 207
Permeability,
- of coal ash, 37
- coefficient of, 33
- intrinsic, 33
- porosity and, 33–34

Phosphogypsum (PG), 49, 74
Phreatomagmatic eruptions, 20
pH value, fly ash, 36
Pieter Bruegel the Elder, 14, 15
Pineapple fiber, 173
Plasticizer, water-soluble, 97, 98, 105, 106, 126, 157, 175, 181
Pollution,
- air, 29–30, 32, 32, 170, 231
- coal-based thermal power stations, 92
- control, 24, 29–30, 42, 47, 72, 228, 234
- environmental, 46, 72, 125, 152, 234
- groundwater, 32
- industrial, 234
- OPC and, 152
- problem, solving, 234

Polypropylene fibers, 173–174
Polysialates, 45
Poly-vinyl alcohol (PVA) fibers, 174
Pond ash (PA), 23, 29–31
- -based GP products, corrosion of. *see* Corrosion, of pond ash (PA)-based geopolymer products
- application areas of, 31
- chemical and physical characteristics, 30

chemical composition of, 30, 95, 126
chemicals,
Sika (water-soluble plasticizer), 97, 98
sodium hydroxide, 97
sodium silicate, 96
collection, 92, 156
development, HCFC slag-based GP. *see* High carbon ferrochrome (HCFC) slag-based GP
engineering properties of, 32
fly ash *vs.*, 30–31
geopolymer production from, 72, 73, 79
macrograph of, 95
management, importance of, 32–33
mechanical activation of, 93
PA/HCFC slag-based GP,
material, characterizations of, 158, 159
mortar and concrete, 158, 159
preparation, 157
phases of, 126
physical properties of, 93, 96
physicochemical properties, 31
preparation of geopolymer from
curing, 100
molding, casting and compaction, 99, 100
raw materials, 98, 99
process of generating, 31
reactivity of, 92, 93
silica content of, 153
storage, 93
variable characteristics of, 32

Pond ash (PA)-based GP cementitious materials,
compressive strength, 104
experimental details, 94–114
materials, 94–98
overview, 92–94
preparation of geopolymer, 98–100
results and discussion, 104–114
test methods, 100–104

Pond ash (PA)-based GPs, strength property of,
experimental details, 126–127
characterization of prepared samples, 127
materials and method, 126
preparation of geopolymer, 126
overview, 124–126
regression coefficients, calculation of, 128, 130–131, 133
results and discussion, 127–147
morphologies of as-received pond ash, 141, 146f
nomograms, 141–147
significance coefficients, testing of, 133–147

Pond ash (PA)–jute fiber-based GP cementitious materials,
experimental details, 175–189
chemicals and materials, 175–178
compressive strength tests, 182–186
concrete preparation, 181
curing process, 179–182
molding, casting and compaction, 179–181
mortar and concrete, 178–185
mortar preparation, 180
preparation, 179
raw material preparation, 178, 179–180, 181
results and discussion, 186–189
test methods, 182–185
microstructures of, 186
overview, 170–175

Pond fly ash (PFA), 31–32

Porosity,
PA-jute fiber-based GP, 174
permeability and, 33–34

Portland cements (PC),
alternatives to, 47–48, 73
hydration characteristics of, 53

OPC. *see* Ordinary Portland Cement (OPC)
Pozzolana, 31
production of, 19, 47, 73
substitute material for, 38
typical composition of fly ash, 28, 29
Portlandite, 49, 75
Potassium aluminate (KA), 156
Potassium metasilicate (KS), 156
Pourbaix diagram, 198, 199
Power-compensated DSC, 103
Pozzolanic effect of ashes, 29–30
Processed pond ash, 178–181
Processed sand, defined, 181
Protective coatings (PCs), geopolymer, 229–230
Protective equipment, use of, 17
Pulverized fuel ash, various uses of, 31–32
Pyrolysis process, 103

Quarry dust, 20
Quartz, 39, 114, 218

Rebar, rusting of, 54
Red mud (RM), 172
Regression coefficients, calculation of, 128, 130–131, 133
Reinforced concrete (RC) building materials, 201
Reinforced fibers, 173
Reinforced pond ash, shear strength of, 171
REMI mold, 99, 126, 179–181, 203–205
Renaissance, the, 11–15
Residual waste glass (RWG), 49, 74
Rheims Cathedral, 9, 10
Rice husk ash (RHA), 19, 172, 173–174
Rock slag, 42
Romanesque style of architecture, 9
Rome, ancient, 5–8
Roofing, 12, 171
Rusting of rebar, 54

Salisbury Cathedral, 15
Sandy silts, 34
Santa Maria del Fiore, in Florence, 12, 13
Sarmin, Siti Noorbaini, 173
Sawdust, 171
Scanning electron microscopy (SEM),
HCFC slag-based GPs, 152, 153, 155, 156, 158–161
PA-based GP mortar/concrete, 206, 214, 217
PA-based GPs, 101–102, 107–110
PA-jute fiber-based GP, 172, 182, 186
zeolite-based geopolymer, 202
Self-cleaning concrete materials, GP-based, 231
SEM. *see* Scanning electron microscopy (SEM)
Sericite powder, 230
Seventeenth century, constructional materials, 15
Shear strength, of coal ash, 37
Shyam Ferro Alloys, 156
Sialate, 45
Significance coefficients, testing of, 133–147
Sika (water-soluble plasticizer), 97, 98, 105, 106, 126, 157, 175, 181
Sikora, S., 200–201
Silica, 19, 151, 153
fume, 93, 174, 202
polymorph, 79
Silica-alumina-bearing waste material. *see* Fly ash (FA)
Silico-aluminates, 45
Silicon-bearing materials, GP preparation, 154
Siloxane oligomers, 79
Silty sands, 34
Sindhunata, 155
Sisal and coir fibers, residual, 171

Skara Brae in Scotland, 2
Slag-based geopolymer concretes (SGPC), 154
Slag(s), 24, 39–45
  ACBFS, 42
  AOD, 48, 73
  BFS, 40–42, 44
  BOF, 42
  Ca electric arc ferronickel, 73, 76
  cement, 42, 44
  chemistry, 39
  converter, 44
  electric arc furnace slag, 44
  ferrous and non-ferrous smelting processes, 39
  GBFS, 42
  generation of, 40–44
  geopolymer production from, 72–78
  GGBFS, 42
  HCFC slag-based GP. *see* high carbon ferrochrome (HCFC) slag-based GP cementitious materials
  iron and steel slag, 43, 44
  LD or steel slag, 42, 44, 45
  metallurgical, 51, 77
  optical photo images, 41
  properties and utilization, 44–45
  reducing, 44
  rock, 42
  sodium sulfate-activated slag cements, 51, 77
  steel. *see* Steel slags
  synthetic, 39
Slave labor, use, 7
Smelting,
  copper, lead and bauxite, 39
  ilmenite, 39
  processes, ferrous and non-ferrous, 39
Soda-lime waste glass powders, 77
Sodium aluminosilicate hydrate, 218
Sodium carbonate, 114, 153
Sodium hydroxide (SH), 49, 50, 74, 76, 97
  from Loba Chemicals, 126
  preparation of geopolymer, 154, 156, 180, 181
  variations of, 104–105
Sodium perborate, 233
Sodium silicate (SS), 48–50, 73–74, 76
  chemical structure of, 96
  from Merck (India), 126, 156–157
  preparation of geopolymer, 154, 156, 180, 181
  variations of, 104–105
Sodium sulfate-activated slag cements, 51, 77
Sol-gel method, for geopolymers, 48, 73
Songyue Pagoda (China), 8
Specific surfaces, of coal ash, 35
Spray coating technique, 200
Sputter coater, 102
St. Paul's Cathedral, England, 15
Stave churches, in Scandinavia, 9
Steel,
  corrosion,
    mechanism of, 54–55
    property of steel-reinforced concrete structures, 55
  dissolution mechanism of, 54
  mass production of, 16
  passivity of embedded, 55
  reinforcement, corrosion of, 198
  use, 16
Steel Age, 3–4
Steel slags, 42
  applications, 44
  compositions, 43, 44
  geopolymer production from, 72, 75
  powder, 50, 75
  production, 40
Strength properties of geopolymers, 48, 51, 53, 74, 77, 93, 104, 175
  at room temperature, 106

compressive. *see* Compressive strength
effect of parameters on, 115
enhancement of, 153
mechanical, 233
mortars, 172
of as prepared composite, 171
of PA-based geopolymer. *see* Pond ash (PA)-based GPs, strength property of
proportion of FA/BFS blends, 153
results and discussion on, 159–163
volume fraction of glass fiber/polypropylene fiber and, 174
Subbituminous coal, 27
Sugar cane bagasse ash (SCBA), 202
Sulfuric acid, 93, 154
Super-sulfated cement (SSC), 49, 74
Supplementary cementitious materials (SCMs), 200
Surface water pollution, 32
Sweet sorghum, 171
Synthetic basalt, 94

Talcum powder, 230
Temple of Apollo at Didyma, 6
Temple of Vesta in Tivoli, Italy, 6
Tensile strength, of geopolymers, 22, 104, 176, 200
Ternary cement mortars (TBCMs), 172
Ternary ecological concrete (TEC), 202
Terracotta, 16
Tetra-calcium aluminoferrite, 44
Thermal conductivity
PA-jute fiber-based GP, 174
TROLIT, 233
Thermal curing, 94
Thermal power plants
ashes, nature and composition of, 24, 26–29
in India, 124
pond ash produced in, 32
Thermal power stations,
coal-based, 92
pond ash and, 153
Thermal stability, 173, 175
by TGA, 127
of as-prepared PA-based GP mortar/concrete, 207
of PA/Jute fiber-based GP, 182
PA/HCFC slag-based GP, 158
zeolite-based geopolymer, 201–202
Thermograms, DSC, 112, 141, 146, 147, 161, 162
Thermogravimetric analysis (TGA),
HCFC slag-based GPs, 158, 159
PA-based GP mortar/concrete, 207
PA-based GPs, 102, 113, 127
PA-jute fiber-based GP, 172
Thermo Nicolet NEXUS 870 FTIR spectrometer, 100
Titanium dioxide, 39
Tower of Babel, 14, 15
Trajan's column in Rome, 8
Treadwheel crane, Roman, 6, 7
Tricalcium silicate, 44
TROLIT, 233
Twentieth century, constructional materials, 17–18

Ultra-high-performance concrete, production, 155
Ultrasonic pulse velocity test, 201
UMT2 Micro Balance, 206
Undrained unconsolidated (UU) triaxial tests, 171
UNESCO World Heritage Site building, 12, 13
United States, coal deposits in, 21, 23
Ur, 4
Uruk, 4

Vegetable fibers, 171
Villard de Honnecourt, 9, 10

Vitruvius, 11–12
Volcanic ash, 19–20

Waste(s),
  due to coal combustion, 92
  fibers in preparation of eco-friendly GP composite, 171–172
  glass-based GPs, 49, 74
  industrial,
    bearing aluminosilicate minerals, 230
    dumping and landfilling, 170
    GPC from, 229
    GP production from, 72, 74, 76, 77
  materials, 229
  molding sands, 73, 76
  PA, dumping and landfilling, 170
  silica-alumina-bearing waste material. *see* Fly ash (FA)
Water impounded hopper (WIH) system, 29
Water-soluble plasticizer (Sika), 97, 98, 105, 106, 126, 157, 175, 181
Weight change, PA-based GP mortar/concrete, 207–213
Weight loss,
  measurement, 200–201
  of bricks, 174
  of GP cured samples, 146, 147
Wet method, 124
Wood ash, 19
Wooden building, Chinese, 8
Wood flour, 171
Wool, 171
Woolworth Building, 16, 17
Wrought iron, use of, 16

X-ray diffraction (XRD), 94
  geopolymer cement using HCFA, 202
  HCFC slag-based GPs, 152, 153, 155, 156, 159
  PA-based GP mortar/concrete, 218
  PA-based GPs, 100, 101, 127
  PA-jute fiber-based GP, 172, 175
  zeolite-based geopolymer, 202
XRD. *see* X-ray diffraction (XRD)

Z300 Benchtop Centrifuge, 206
Zeolite-based geopolymer, 76, 80, 201–202, 218
Ziggurat of Ur, 4

Printed and bound by CPI Group (UK) Ltd, Croydon, CR0 4YY
17/09/2026
14964277-0001